COLORADO GEOLOGICAL SURVEY
BOULDER

R. D. GEORGE, State Geologist

BULLETIN 30

GEOLOGY AND ORE DEPOSITS

OF THE

RED CLIFF DISTRICT

COLORADO

By

R. D. CRAWFORD and RUSSELL GIBSON

BOULDER, COLORADO
CAMERA PRINT
1925

GEOLOGICAL BOARD

LETTER OF TRANSMITTAL

STATE GEOLOGICAL SURVEY,

UNIVERSITY OF COLORADO, JULY 8, 1924

Governor William E. Sweet, Chairman, and Members of the Advisory Board of the State Geological Survey.

GENTLEMEN: I have the honor to transmit herewith Bulletin 30 of the Colorado Geological Survey.

Very respectfully,

R. D. GEORGE,

State Geologist.

CONTENTS

Chapter VI (Continued)

Chapter VII (Continued)

ILLUSTRATIONS

GEOLOGY AND ORE DEPOSITS

OF THE

RED CLIFF DISTRICT, COLORADO

INTRODUCTION

GEOGRAPHIC POSITION

The Red Cliff, or Battle Mountain, mining district is in the southeastern part of Eagle County, Colorado, west of the Continental Divide. An area of about 58 square miles has been mapped, and is described in this report. Gilman, Bell's Camp, and Red Cliff are in the productive part of the district. The main line of the Denver and Rio Grande Western Railroad

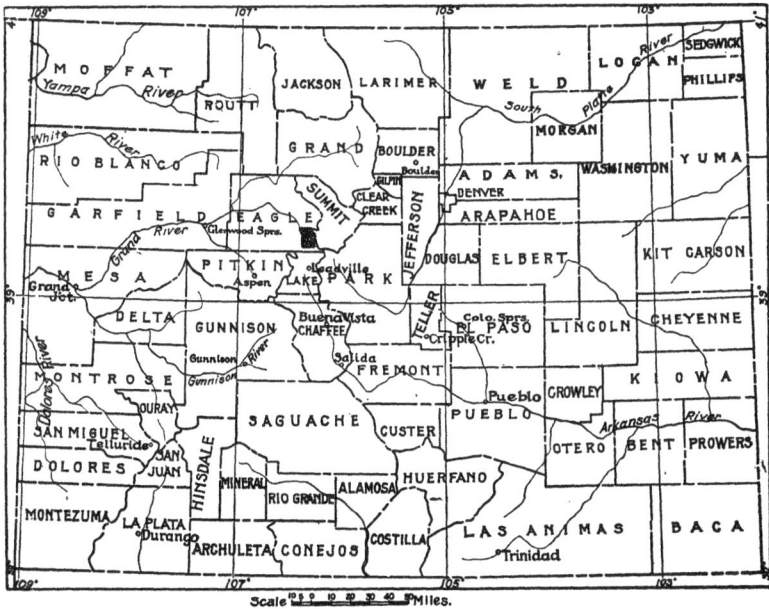

Figure 1. Index map of Colorado, showing position of the Red Cliff district

passes through all the present settlements of the district except Gilman and Bell's Camp. These two are near the railroad, but high above the bottom of Eagle Canyon through which the railroad runs. Gilman is connected with Belden, at the bottom of the canyon, by a tramway of the Empire Zinc Company. The Pikes Peak Ocean to Ocean Highway runs through Pando, Red Cliff, Bell's Camp, and Gilman.

FIELD WORK AND ACKNOWLEDGMENTS

The field work for this report was done in the summers of 1921, 1922, and 1923 by a party that varied in number from two to five. Each man regularly employed spent from two to six months in the field. In the productive area and much of the outlying territory topography and geology were mapped at the same time. In this work a plane-table and telescopic alidade were used, and distances and elevations determined by stadia. One instrument man platted on a single plane-table sheet the data obtained by two or three rodmen, together with the data obtained by himself at and near the plane-table stations. Each rodman platted on tracing paper the topography, geology, and cultural features at and near the stadia stations. These data were transferred to the plane-table sheet at convenient intervals,—at the end of the day when instrument man and rodman were working on opposite sides of a canyon. In many places timber or high relief made extensive linear traverses impracticable; in consequence, sights were taken to many stadia stations from a single plane-

Figure 2. Red Cliff, looking toward Turkey Creek

table station. Part of the area was mapped topographically by means of paced traverses with traverse plane-table, open-sight alidade, and aneroid barometer. The entire map was controlled by triangulation with plane-table and telescopic alidade, beginning at the ends of a measured base line in Eagle Park. The control was tied to Holy Cross and Homestake mountains whose latitude and longitude had been determined by the United States Geological Survey. A large-scale topographic map of part of Battle Mountain made by Messrs. Platt and Kleff for the Empire Zinc Company, was reduced, and is included in the accompanying map (Pl. I, in pocket).

The personnel of the field party changed each season. The topographic map and the greater part of the geologic map were made by H. E. Alexander, E. P. Andrews, H. C. Fuller, Russell Gibson, E. A. Hall, J. W. Hunter, J. E. Hupp, A. N. Murray, and W. O. Thompson. Mr. Gibson was in the field about six months, and spent most of his time in geologic work. R. D. Crawford was in the district about ten weeks, and was engaged in geologic work exclusively. Most of the necessary laboratory work, except the fossil determinations and quantitative analyses, has been done by Mr. Hall and the writers, who, in preparing this report, have used the notes of all the field men.

The writers gratefully acknowledge the uniformly generous spirit of cooperation shown by mine operators, miners, and others, many of whom gave much time to further the survey of the district. Thanks are especially due Messrs. R. B. Paul and A. H. Buck of the Empire Zinc Company, as well as other members of their staff, for their many courtesies; Messrs. J. M. and R. V. Dismant and Mr. B. A. Hart for maps, information concerning production, and other assistance; Mr. Arthur Ridgway of the Denver and Rio Grande Western Railroad for a large-scale map of the right-of-way of the railroad, and Mr. C. W. Henderson of the United States Geological Survey for the table (p. 12) showing the production of the district in detail.

Figure 3. Gilman from the southeast

HISTORY OF MINING

Ore was discovered in the Gilman and Red Cliff region in 1879 when it was part of Summit County. First known as the Eagle River mining district the area has in recent years been called the Red Cliff district or, possibly more often, the Battle Mountain mining district. Eagle County was organized in 1883.

A series of well written and interesting articles by W. B. Thom, published by "The Eagle Valley Enterprise," of Eagle, in 1921, gives much information concerning the early history of mining and the mining camps of Eagle County. Mr. Thom states that among the earliest locations on Battle Mountain, mostly in 1879, "were the Little Ollie, by James Deming, William

Barney, and William Helmer; the Eagle Bird, by J. T. McGrew, William and Henry Helmer and D. C. Collier; the Silver Wave, by Helmer brothers; the Black Iron, by Dr. Bell and B. S. Morgan. Other properties were the Ida May, Little Duke, May Queen, Kingfisher, Little Chief and Crown Point. Robert Haney and William Travis feathered their nests from the Ben Butler. In 1880 A. W. Maxfield represented Council Bluffs mine owners on Battle Mountain. He was followed by J. T. Hart of Council Bluffs. . . . . Among the first locations at Red Cliff were the Henrietta, by Frank Bowland & Co., July 4, 1879; the Horn Silver, by Wm. Greiner and G. J. DaLee, July 7, 1879, and the Wyoming, by Dugan, Jenkins and others, July 26, 1879. . . . . "

The foregoing list of mining claims includes some that afterward became heavy producers. According to Mr. Thom the first important ore discovery was made in the Belden mine, located May 5, 1879. He also states that the Battle Mountain smelter was "built in 1880, only to be abandoned when the Denver and Rio Grande Railway entered Red Cliff, November 20, 1881."

Figure 4. Gilman from the northwest

. In the 80s much development work was done at Taylor Hill, a short distance south of the area mapped, and at Holy Cross City, a few miles west of the Red Cliff district. Each of these localities produced several thousand dollars worth of ore, but the returns were small in proportion to the invested capital. Near Two Elk Creek, a short distance north of the mapped area, some prospecting has been done, and ore minerals have been found.

The metal production of Eagle County by years is shown in Table I. This table was prepared by C. W. Henderson, Mineral Geographer of the United States Geological Survey, for his "History of Mining in Colorado," and was generously furnished by Mr. Henderson for this bulletin. According to Mr. Henderson the Red Cliff district should be credited with all this production excepting 200,000 ounces of silver produced by the Brush Creek district, in the years 1913-1918, and the several thousand dollars worth of ore produced in the early days by the Taylor Hill and Holy Cross localities. At the end of 1923 the Red Cliff district had therefore produced ore to the value of $28,000,000 or more. The average values of the metals for each year are shown in Table II which, down to and including 1922, is taken from a larger table in "Mineral Resources of the United States, 1922," published by the United States Geological Survey; Mr. Henderson kindly supplied the average values for 1923.

Several mills have been built in the district, and formerly much ore was concentrated in the Iron Mask mill at Belden. This mill has been idle for several years, and nearly or quite all the ore mined in recent years has been shipped directly from the mines—part to the smelters, and part to the Empire Zinc Company's plant at Canon City, Colorado.

BIBLIOGRAPHY

Peale, A. C., Report of geologist of Middle Division: Hayden Survey, 8th Annual Report, 1876, pp. 73-180.

Bechler, G. R., Geographical report on the Middle and South parks, Colorado, and adjacent country: Hayden Survey, 9th Annual Report, 1877, pp. 371-440.

Emmons, S. F., Notes on some Colorado ore deposits: Colorado Scientific Society Proceedings, vol. 2, 1886, pp. 85-105.

Tilden, G. C., Mining notes from Eagle County—The quartzite deposits of Red Cliff; Taylor, Eagle County: Colo. School of Mines, Biennial Report, 1886, pp. 129-133.

Olcutt, E. E., Battle Mountain mining district, Eagle County, Colorado: Engineering and Mining Journal, vol. 43, 1887, pp. 418-419, 436-437.

Guiterman, F., Gold deposits in the quartzite formation of Battle Mountain, Colorado: Colorado Scientific Society Proceedings, vol. 3, 1890, pp. 264-268.

Rosenberg, Leo von, The mines of Battle Mountain, Eagle County, Colorado: Engineering and Mining Journal, vol. 53, 1892, p. 545, cross-section on p. 544.

Lakes, Arthur, Redcliff ore deposits; Mines and Minerals, vol. 23, 1903, pp. 252-253.

Lakes, Arthur, The Red Cliff ore deposits: Geology of Western Ore Deposits, 1905, pp. 242-244.

Nicholson, H. H., Low grade sulphides of Eagle County, Colorado: Mining Science, vol. 64, 1911, pp. 127-128.

Hoskin, A. J., Revival of mining at Red Cliff: Mines and Minerals, vol. 33, 1912, pp. 147-151.

Means, A. H., Geology and ore deposits of Red Cliff, Colorado: Economic Geology, vol. 10, 1915, pp. 1-27.

TABLE I—TOTAL PRODUCTION OF GOLD, SILVER, COPPER, LEAD, AND

Year	Short Tons Ore	Gold Fine Ounces	Gold Value	Silver Fine Ounces	Silver Value	Recoverable Copper, Pounds
a1879						
1880	c 1,000	....	c $ 2,000	b 38,672	$ 44,473	....
d1881	c 2,000	....	c 4,000	c 79,344	89,659	....
e1882	c 2,500	....	c 5,000	c 98,680	112,495	....
1883	g 19,859	....	f 70,000	f 232,031	257,554	....
1884	ch 10,000	....	h 30,000	h 154,687	171,703	....
1885	j 11,000	....	k 33,000	k 170,156	182,067	....
1886	....	....	l 423,517	l 569,637	563,941	....
1887			l 219,594	l 254,078	248,996	
1888			l 142,002	l 193,489	181,880	
1889			l 92,220	l 170,551	160,318	
1890			l 68,862	l 75,265	79,028	
1891			l 153,453	l 280,168	277,366	
1892			l 139,299	l 347,954	302,720	
1893			l 168,867	l 187,658	146,373	
1894			l 55,521	l 62,543	39,402	
1895			l 30,900	l 53,421	34,724	
1896			l 16,472	l 65,824	44,760	l 2,044
1897			m 34,767	m 46,046	27,628	m 2,200
1898			m 30,571	m 70,783	41,762	m 71,049
1899			m 46,094	m 44,393	26,636	m 5,876
1900			m 103,598	m 234,674	145,498	m 359,054
1901			m 97,376	m 175,181	105,109	m 157,914
1902			m 31,956	m 45,336	24,028	m 150,134
1903			m 16,040	m 27,054	14,609	m 32,863
1904	n 1,866		m 30,075	m 27,348	15,862	m 32,409
1905	n 12,049		n 46,891	n 46,487	28,357	n 29,331
1906	n 15,986		n 51,561	n 94,912	64,540	n 130,233
1907	n 4,191		n 53,641	n 70,586	46,587	n 14,270
1908	n 3,009		n 58,131	n 86,715	45,959	n 66,141
1909	n 11,526		n 53,308	n 125,214	65,111	n 286,885
1910	n 27,761		n 25,231	n 88,313	47,689	n 209,551
1911	n 33,177		n 41,160	n 116,109	61,538	n 66,608
1912	n 34,164		n 49,294	n 163,735	100,697	n 147,176
1913	n 47,488		n 41,220	n 301,380	182,034	n 41,368
1914	n 49,377		n 47,194	n 127,080	70,275	n 28,105
1915	n 74,197		n 95,426	n 177,550	90,018	n 60,086
1916	105,149	4,645.74	96,036	222,126	146,159	112,610
1917	100,875	1,992.42	41,187	136,023	112,083	53,136
1918	89,675	1,740.29	35,975	241,406	241,406	353,041
1919	22,248	964.36	19,935	72,159	80,818	123,306
1920	32,635	1,233.37	25,496	279,667	304,837	517,109
1921	39,785	3,130.98	64,723	682,550	682,550	1,833,078
1922	71,892	3,488.37	72,111	583,737	583,737	1,330,296
1923	....	2,060.69	42,598	327,593	268,626	507,612
			$3,06,302	7,648,315	$6,561,612	6,723,485

FOOTNOTES FOR TABLE I

By CHARLES W. HENDERSON

a Director of the Mint Report for 1883, pp. 240, 290-294 :—

"Eagle County was formerly a part of Summit, and embraces what were known as Eagle River and Holy Cross mining districts. . . . . The chief camp is Red Cliff, in the vicinity of which are the largest producers in the county, viz., the Eagle Bird group, Belden, Clinton, Little Chief, Crown Point, Spirit, King Fisher, Discovery, Silver Age, and a few minor properties. These are located on Battle Mountain. . . . The Belden [was] located in 1879. . . . Eagle Bird Consolidated Mining Company. . . . consists of 25 claims, among which are . . . . the Silver Wave, Eagle Bird, Indian Girl, May Queen, Cleveland, Black Iron. . . ."

1880

b Director of the Mint for 1880, p. 152, gives, under Summit County, $50,000 silver (38,672 ounces at $1.29+) as production of Eagle River district. . . . "Several promising contacts, notably the Belden group, have been opened the past summer."

1880-1882

c Estimated by Chas. W. Henderson.

ZINC IN EAGLE COUNTY, COLORADO, 1879-1923

Copper Value	Recoverable Lead		Recoverable Zinc		Total Value	Year
	Pounds	Value	Pounds	Value		
....			....	....	$ 86,473	1879
....	c 800,000	$ 40,000	....	....	170,459	1880
....	c 1,600,000	76,800	....	....	215,495	1881
....	c 2,000,000	98,000	....	....	1,015,554	1882
....	g 16,000,000	688,000	....	....	445,903	1883
....	i 6,600,000	244,200	....	....	447,117	1884
....	j 5,950,000	232,050	....	....	1,079,458	1885
....	cl 2,000,000	92,000	....	....	518,671	1886
....	l 1,112,905	50,081	....	....	428,166	1887
....	l 2,370,090	104,284	....	....	334,917	1888
....	l 2,112,280	82,379	....	....	192,890	1889
....	l 1,000,000	45,000	....	....	593,197	1890
....	l 3,776,230	162,378	....	....	652,390	1891
....	l 5,259,280	210,371	....	....	500,240	1892
....	cl 5,000,000	185,000	....	....	160,923	1893
....	cl 2,000,000	66,000	....	....	122,271	1894
....	l- 1,770,215	56,647	c	....	67,775	1895
$ 221	l 210,717	6,322	....	....	103,843	1896
264	m 1,144,013	41,184	....	....	151,500	1897
8,810	m 1,851,512	70,357	....	....	127,192	1898
1,005	m 1,187,930	53,457	....	....	471,491	1899
59,603	m 3,679,328	161,912	cm 20,000	$ 880	348,195	1900
26,372	m 2,775,291	119,338	....	....	108,447	1901
18,316	m 832,846	34,147	....	....	63,616	1902
4,502	m 677,730	28,465	....	....	66,219	1903
4,148	m 375,207	16,134	....	....	122,921	1904
4,576	n 156,723	7,366	m 605,612	35,731	245,766	1905
25,135	n 307,755	17,542	m 1,426,029	86,988	138,671	1906
2,854	n 193,690	10,266	m 429,198	25,323	113,292	1907
8,731	n 11,204	471	....	....	202,244	1908
37,295	n 152,280	6,548	n 740,408	39,982	341,008	1909
26,613	n 397,409	17,486	n 4,147,945	223,989	440,102	1910
8,326	n 855,889	38,515	n 5,097,597	290,563	620,571	1911
24,284	n 1,240,156	55,807	n 5,659,261	390,489	663,403	1912
6,412	n 1,351,205	59,453	n 6,683,643	374,284	550,752	1913
3,738	n 1,177,385	45,918	n 7,522,098	383,627	1,643,056	1914
10,515	n 1,394,043	65,520	n 11,141,750	1,381,577	4,185,294	1915
27,702	1,517,362	104,698	28,438,052	3,810,699	2,795,469	1916
14,506	2,426,988	208,721	23,715,412	2,418,972	1,923,332	1917
87,201	2,927,099	207,824	14,845,341	1,350,926	389,559	1918
22,935	378,113	20,040	3,367,548	245,831	986,996	1919
95,148	282,538	22,603	6,653,235	538,912	984,306	1920
236,467	12,578	566	....	....	1,480,193	1921
179,590	322,818	17,755	11,000,000	627,000	2,044,585	1922
35,533	632,846	93,028	23,600,000	1,604,800		1923
$980,802	87,824,155	$3,964,633	155,093,129	$13,830,573	$28,343,922	

1881

d Director of the Mint for 1881, p. 435, under Summit County, says: "On Eagle River, at Red Cliff, the Belden Mining Company and the Battle Mountain smelter have produced a large quantity of base bullion. . . . No ore is being shipped except from the Belden Company's mine, and from this but four car-loads per day. The ore, containing as it does a large percentage of lead, forms an excellent flux for the dry ores and will aid materially in smelting. In Gold Park, near the Mount of the Holy Cross, a considerable camp has been established, and a large stamp-mill erected to crush the gold ores found there. The veins are quite strong, the quartz at the surface to a depth of 12 to 15 feet, much decomposed, and the yield $10 to $25 per ton under stamps. Below the decomposition the ores are mainly white iron pyrites. This section is scarcely more than a year old. . . . ."

1882

e Director of the Mint for 1882, p. 559, under Summit County, says, "At Red Cliff, the Belden is the principal mine."

1883

f Director of the Mint for 1883, p. 240.
g Idem, p. 294.

1884

h Director of the Mint for 1884, pp. 177, 207-209:—
". . . . The Belden shipped during the year 1,640 tons of ore, which returned as follows:

Ores	Quantity	Price	Value
Lead, tons _____	568	$72.00	$40,896
Silver, ounces _____	16,400	1.10	18,040
Gold, ounces _____	21½	20.00	430
			59,366

". . . . The Eagle Bird shipped 2,090 tons of ore, assaying 31 per cent lead and 8½ ounces silver, giving 647 tons of lead and 17,765 ounces of silver. Black Iron produced 378 tons of ore, assaying 41 per cent lead and 9 ounces silver, giving 154.98 tons lead and 3,402 ounces silver. The May Queen forwarded 30 tons which returned 39 per cent lead and 8 ounces silver, giving 11 tons lead and 240 ounces silver. The total of these mines comprising the Cheeseman and Clayton group was as follows:

Ores	Quantity	Price	Value
Lead, tons _____	814.58	$72.00	$58,659.76
Silver, ounces _____	21,407	1.10	23,547.70
Gold, ounces _____	157	20.00	3,140.00
			85,347.46

The shipments from the Little Chief, Crown Point, Kingfisher, Iron Mask, Spirit, Clinton, Cleopatra, Great Western, and Potvin have amounted to some 597 car-loads of mineral, averaging 10 tons to the car; 32 per cent lead and 8 ounces silver per ton would give 5,970 tons of mineral yielding as follows:

Ores	Quantity	Price	Value
Lead, tons _____	1,910.4	$72.00	$137,548.88
Silver, ounces _____	4,776	1.10	52,536.00
Gold, ounces _____	250	20.00	5,000.00
			195,084.88

On the lower end of Battle Mountain is found what is now termed the Quartzite, which extends on the west side of Rock Creek and on the north side of Eagle River. The most promising of its mines are the Ground Hog group, Combined Discovery, Uncle Sam, Horn Silver and Highland Mary. . . . The total yield of Red Cliff Camp has been stated to be, from the Carbonate mines, $339,798.26; from other mines, $125,000; total, $464,798.26 (which includes the value of lead).
"In the vicinity of Taylor Hill much work is being done in the way af development, and McClelland's stamp-mill has been running most of the year on ore from this locality. It is claimed about $12,000 was turned out during this time. At Holy Cross exploitation only has been done; bodies of ore have been developed, but the actual yield has been almost nothing. The production of the camp, reported from ore treated, was $6,400."

1884

i Mineral Resources for 1884, p. 422.

1885

j Mineral Resources for 1885, p. 257, gives 11,000 tons of ore containing 3,500 tons lead gross; deducting 15 per cent gives 2,975 tons lead.
k Interpolated by Chas. W. Henderson, to correspond with total production of the state.

1886-1896

l From reports of the agents of the Mint, in annual reports of the Director of the Mint, prorating the gold and silver to correspond with the corrected figures of the total production of the State by the Director of the Mint, and prorating the lead production to correspond with the total production of lead for the State in annual volumes of Mineral Resources, and distributing any "unknown production" of the State proportionately to the various counties. Similarly with copper as with lead, but as Mineral Resources figures for copper include copper from matte and ores treated in Colorado smelters from other states, the copper figures are subject to revision.

1897-1907

m Colorado State Bureau of Mines figures being smelter and Mint receipts.

1905-1918

n Mineral Resources of the United States, U. S. Geological Survey, figures collected from the mines.

TABLE II.   PRICES OF SILVER, COPPER, LEAD AND ZINC, 1880-1923

Year	Silver Per Fine Ounce	Copper Per Pound	Lead Per Pound	Zinc Per Pound	Year	Silver Per Fine Ounce	Copper Per Pound	Lead Per Pound	Zinc Per Pound
1880	$1.15	$0.214	$0.05	$0.055	1902	$0.53	$0.122	$0.041	$0.048
1881	1.13	.182	.048	.052	1903	.54	.137	.042	.054
1882	1.14	.191	.049	.053	1904	.58	.128	.043	.051
1883	1.11	.165	.043	.045	1905	.61	.156	.047	.059
1884	1.11	.13	.037	.044	1906	.68	.193	.057	.061
1885	1.07	.108	.040	.043	1907	.66	.20	.053	.059
1886	.99	.111	.046	.044	1908	.53	.132	.042	.047
1887	.98	.138	.045	.046	1909	.52	.13	.043	.054
1888	.94	.168	.044	.049	1910	.54	.127	.044	.054
1889	.94	.135	.039	.05	1911	.53	.125	.045	.057
1890	1.05	.156	.045	.055	1912	.615	.165	.045	.069
1891	.99	.128	.043	.05	1913	.604	.155	.044	.056
1892	.87	.116	.041	.046	1914	.553	.133	.039	.051
1893	.78	.108	.037	.04	1915	.507	.175	.047	.124
1894	.63	.095	.033	.035	1916	.658	.246	.069	.134
1895	.65	.107	.032	.036	1917	.824	.273	.086	.102
1896	.68	.108	.03	.039	1918	1.00	.247	.071	.091
1897	.60	.12	.036	.041	1919	1.12	.186	.053	.073
1898	.59	.124	.033	.046	1920	1.09	.184	.08	.081
1899	.60	.171	.045	.058	1921	1.00	.129	.045	.05
1900	.62	.166	.044	.044	1922	1.00	.135	.055	.057
1901	.60	.167	.043	.041	1923	.82	.147	.07	.068

CHAPTER I
PHYSIOGRAPHY
By R. D. Crawford

TOPOGRAPHY

The Red Cliff district is in a region of steep slopes and deep gulches and canyons. The vertical range within the area mapped is approximately 3,500 feet—from about 8,400 to 11,900 feet above sea level. A few miles southwest is Holy Cross Mountain with an altitude of nearly 14,000 feet. The district is drained by Eagle River and its branches, several of which carry large volumes of water throughout the summer. Homestake Creek, which joins Eagle River below Red Cliff, is longer and probably carries more water than that part of Eagle River above the mouth of Homestake Creek. This creek has an advantage in heading near the top of the Sawatch Range where glacial lakes are numerous and snow is found almost or quite throughout the year. Aside from ordinary stream erosion factors that have had part in shaping the relief include geologic structure and character of the rocks, glaciation, landslides, and vegetation.

INFLUENCE OF STRUCTURE

By reference to the geologic map (Pl. I, in pocket) it will be seen that Eagle River flows in general nearly in the direction of strike of the Sawatch quartzite, but a little to the right of this direction, that is, inclining slightly toward the dip of the beds. West of the river the uniformity of slope is well shown by the contours between Pando and Camp Bryan on the map, and is very striking when seen from points in the field. (See fig. 5.) This slope is nearly parallel to the bedding planes of the quartzite which extends about two miles west of the river. From this slope the soft beds of the Sawatch formation and all overlying beds have been removed. Beyond the border of the quartzite the slope of the surface of pre-Cambrian rocks is similar. This uniformity of slope suggests planation by lateral migration of the stream to its present position. A few local bends eastward may have been brought about by greater erosion of rocks weakened by faulting, with or without subsequent mineralizaton. The positions of several gullies in the quartzite have been determined by faults. While it is probable that some of the larger gullies in the higher formation were so determined, efforts to prove this relationship have been unsuccessful. The fault nearly midway between Pando and Red Cliff (Pl. I) strikes southwest toward Homestake Gulch, and may have determined the position of the gulch which is notably straight for a distance of 8 or 9 miles southwestward.

GLACIATION

Within the area mapped glacial boulders are very common on both sides of Eagle River, and thick moraines cover the bedrock in many places. The positions of the largest morainal patches are shown on the geologic map. East of the river the moraines are mostly less than 800 feet vertically from the valley bottom; west of the river glacial boulders are found at practically all elevations between the river and the cirques near the top of the Sawatch Range. Many glacial lakes are found between the valley and the crest of the range. Near Pando rocks show smoothing by moving ice, and roches moutonnees are very common in Homestake Gulch.

Two large glaciers have moved through the region, and converged or followed nearly the same path at and from a point about two miles northwest of Pando. One started near Fremont Pass in the Tenmile district, moved westward north of Chalk Mountain, and down the valley of Eagle River. There is good evidence that the greater part of the ice moved northwestward through the valley now followed by the highway north of Pando (Fig. 5), and not through the present valley of Eagle River where it bends to the right between Coal and Silver creeks. However, nearly

Figure 5. Glaciated valley connecting Eagle and Homestake valleys, looking northwest; shows also dip slope of Sawatch quartzite in background

all the hill between Eagle River and Homestake Creek has been glaciated, and considerable morainal material is found close to Eagle River south of Red Cliff. A larger glacier headed high on the slopes of the Sawatch Range, moved down through Homestake Gulch, and scoured out the gulch to a depth of about 1,000 feet and to a width of a quarter of a mile. Midway between Pando and Red Cliff it joined the Eagle glacier, or followed the path of the Eagle glacier from this point northward. Other glaciers came into Eagle Valley north of the Homestake glacier. Both the Homestake and Eagle glaciers eroded their valleys to a depth considerably below the pre-glacial level. This is shown by the many hanging valleys, of which one is Yoder Gulch where the creek forms a waterfall near the railroad (Fig. 6). Eagle Canyon between Homestake Creek and Rock Creek has evidently been much deepened by stream erosion since the glaciers melted. (See also interpretation by M. R. Campbell in Bulletin 707, U. S. Geological Survey, p. 115.)

LANDSLIDES

Several landslides are found east of Eagle River on the escarpment slope, and a few of the largest are represented on the geologic map. Faults can be seen in the quartzite opposite some of these landslides; it is probable that steep-walled gullies formed in the limestone and shale along the same faults east of the stream, thus facilitating the slump of the shale cliffs. One of the largest areas covered by a landslide is south of Silver Creek. In this vicinity Eagle River evidently was formerly east

Figure 6. A hanging valley, Yoder Falls

of its present position, and gradually undercut the shale beds until they formed a steep cliff which slid down and forced the stream westward. About half a mile north of the ranger station the stream was dammed by the landslide, and the slow-running ponded water has earned the local name Stillwater for this part of the stream and its surroundings. Between the south border of the slump dam and the mouth of Silver Creek the stream is still flowing in rapids over its new bed of surficial material (Fig. 7.)

VEGETATION

Outcrops of the more resistant rocks and steep escarpment slopes above landslides are practically the only surfaces free from grass or trees. Great volumes of water from heavy rains or cloudbursts carry large quantities of mud and boulders down the streams, but the vegetation prevents general rapid erosion during the season when the ground is not covered with snow. When the streams are fed chiefly by springs and melting snow the water is clear. Many steep escarpment slopes and others are covered by grass or timber, excepting a few bare outcrops of limestone and sandstone. The trees are principally pine, spruce, fir and aspen. Much of the timber has been cut for lumber, stulls, and lagging; but there still remain many trees large enough to cut for lumber and mine timber.

Figure 7. Eagle River, looking south from near Silver Creek
The stream has been forced westward by a geologically recent landslide over which the water still flows in rapids. Note in the left background the trees that have grown since the surface rock slid to its present position. The area covered by this landslide is shown on the geologic map in pocket.

ALLUVIUM AND SOIL

Owing to the generally high gradient of the streams and the steep valley walls a relatively small amount of alluvium has accumulated. A few alluvial cones and areas of comparatively flat grass-covered alluvium, in part a re-worked glacial deposit, are found in the Eagle and Homestake valleys. Though the soil of these flats is fertile the growing season is short and the nights are cold. Consequently the land has been used chiefly for pastures and meadows; a few vegetables have been raised for home use. In 1923 head lettuce for outside markets was successfully grown in Eagle Valley in the southeast part of Eagle Park.

CHAPTER II

PRE-CAMBRIAN METAMORPHIC ROCKS

By R. D. CRAWFORD

The oldest rocks in the region include gneiss, schist, quartzite, and marble. The gneiss and schist alternate with each other and intergrade; within the gneiss and schist area as small patches of quartzite and marble. No attempt has been made to map any one of the kinds of metamorphic rocks separately.

Near the border of certain gneiss areas granite, in streaks, has been injected along the foliation planes, forming either a small or a large proportion of the rock volume. These relationships of the two kinds of rock make impossible an exact placing of the boundary between them. Where these conditions are found the aim has been to represent on the map as granite the rock that is dominantly granite, and to represent as gneiss and schist those areas in which gneiss and schist form more than half the rock mass. The same procedure was followed where the granite was intruded as dikes or small stocks in the gneiss and schist, and where the granite incloses large masses of the older rocks. Good illustrations of granite intrusions along the structure planes of the older rocks may be seen near the mouth of Homestake Creek. Farther south, east of the same creek, granite cuts across the foliation planes as dikes and stocks.

The foliation planes of the gneiss and schist strike in a general northeast direction, and are usually nearly vertical or dip northwestward at a high angle. In general there has not been intense folding, though locally the bands are considerably contorted.

Part of the gneiss is evidently of igneous origin. In rock of this type the gneissic structure may be in part a fluxional arrangement of the minerals, but the structure was probably more generally imposed by pressure during the process of metamorphism. The high proportion of quartz and the comparatively low feldspar content, with not infrequently high mica content, indicate that some of the gneisses and schists are metamorphosed sediments. This is to be expected in a region that contains the metamorphic equivalents of sandstone and limestone. Though no very unusual types of metamorphic rocks have been observed pre-Cambrian marble—found here in two localities—seems to be rare in Colorado.[1] Varieties of the metamorphic rocks examined in thin sections and described below should be considered only as types; intergradations are common.

GRANITE GNEISS

Granite gneiss is here taken as a gneissic rock of igneous origin and granitic composition whether the structure was effected by flow when the magma was viscous or by metamorphic processes after solidification. Good examples may be seen a mile west of Red Cliff and about 1.5 miles south-

---

[1] Doctor Cross has reported an occurrence of marble with schist near Tincup, Gunnison County, Colorado, in the Proceedings of the Colorado Scientific Society, vol. 4, 1893, pp. 286-293.

east of Pando. The gneiss varies from coarse to fine grain in different localities. Essential component minerals are orthoclase, microcline, quartz, and biotite.

QUARTZ-DIORITE GNEISS

A small exposure of quartz-diorite gneiss is found west of the mouth of Homestake Creek. The rock has a coarse texture and rough gneissic structure. Probably half the rock mass is composed of hornblende and biotite in nearly equal quantities. The feldspar, insofar as determinable, is plagioclase. Quartz forms about 15 per cent of the rock.

GRANITIC GNEISS

Gneiss that resembles granite in mineral composition, and of uncertain or mixed origin, may be called granitic gneiss. One variety is pinkish-gray, and composed chiefly of quartz and feldspar. It also carries an appreciable quantity of muscovite and very little biotite. The rock may be a metamorphosed arkose or feldspathic sandstone. The proportion of feldspar is a little low for typical granite. Another variety bears a resemblance to the biotite gneiss described below, but contains a higher proportion of injected igneous material. A phase of this rock carries much soda-lime feldspar, and approaches granodiorite in mineral composition.

MICA GNEISS

The mica gneiss differs from the granitic gneiss chiefly in the much higher proportion of mica or quartz or both. The feldspar content is low. Part or all of the rock here called mica gneiss is of sedimentary origin. The rock is dark gray or reddish-gray, and has pronounced gneissic structure. Quartz in most specimens makes up 50 to 80 per cent of the rock. The highly quartzose variety might be called a quartzite, but the pronounced banding is more characteristic of gneiss. Biotite composes about 15 to 30 per cent of the rock mass. Muscovite is present in some specimens.

PYROXENE GNEISS

North of Yoder Gulch, about 200 feet above the railroad, is a small exposure of fine-grained pyroxene gneiss. The rock is dark in color, and contains thin veinlike streaks and lenslike patches of quartz. The most abundant mineral is pyroxene determinable only in thin section. It is grass-green to greenish-gray, monoclinic, and has a birefringence of about .028 or .029. It is probably ferriferous diopside or common pyroxene. Quartz, orthoclase, microcline, soda-lime feldspar, and a little titanite and zircon are present.

SILLIMANITE GNEISS

South of Eagle River, west of the mouth of Homestake Creek, is a small exposure of peculiar rock whose weathered surface shows many bluish patches of a mineral with a splintery fracture. This mineral is shown by the microscope and chemical tests to be sillimanite. Other minerals present are quartz, andesine, and biotite. Any one of the minerals may be bunched, and associated with smaller amounts of the others. That part of the rock rich in sillimanite is about one-half sillimanite in grains up to three-fourths of an inch in diameter. The rock is probably metamorphosed clay or shale with injected igneous material.

MICA SCHIST

The mica schist differs from the mica gneiss described in that it is finely laminated. The mica, forming one-third to one-half the rock, is mostly biotite, though muscovite is sometimes seen. Quartz is the dominant mineral in some specimens; in others, potash feldspar or soda-lime feldspar may exceed the quartz. Magnetite and apatite are present in small quantity. Near the south border of the mapped area is a small exposure of a peculiar mica schist with microcrystalline texture. It shows blocky jointing, and breaks with a conchoidal fracture. It is composed essentially of biotite and quartz.

The highly quartzose varieties of mica schist were probably derived from sediments, but the high feldspar content of some specimens indicates a possible igneous origin for part of the schist.

GARNET-MICA SCHIST

West of the railroad, less than a mile south of Deen Station, is found a garnet-bearing mica schist. Biotite makes up about three-fourths of the rock. Garnet and quartz are the only other essential components of the specimen examined. The garnet is red, and is probably almandite. It is in roundish anhedral grains that inclose microcrystalline grains of quartz and flakes of biotite.

QUARTZITE

Small patches of quartzite have been found in several places between Red Cliff and Pando. Quartzite composed almost entirely of quartz grains is in contact with marble high on the slope west of Homestake Creek, about 1.5 miles south of Red Cliff. A micaceous quartzite is found near Eagle River less than 2 miles south of Red Cliff. About 10 to 15 per cent of this rock is biotite; the rest is mostly quartz. About a quarter of a mile south of the mouth of Homestake Creek is a small outcrop of very fine-grained greenish-gray pyrite-bearing quartzite. As seen under the microscope about half of the rock is quartz in irregular grains; the rest is chiefly epidote in minute grains. The rock was probably once a calcareous sandstone.

MARBLE

Two different kinds of marble are found in the district: (1) a dense variety, finer in texture than much of the Paleozoic limestone and dolomite of the region; (2) a normal granular variety.

The first is found in a lenticular mass 4 or 5 feet thick about half a mile S. 10° E. of the mouth of Rule Creek. The marble is inclosed by gneiss, and can be traced along the strike 100 feet or more. Traces of the bedding of the original limestone are preserved on the weathered surface by small nodular masses and individual grains of resistant minerals. On fresh surfaces the marble is bluish-gray. About 80 per cent of the rock is soluble calcite with practically no magnesium. In thin section the microgranular calcite shows very few traces of cleavage and no twinning. Other minerals scattered throughout the slide in small grains are quartz, orthoclase, plagioclase, and monoclinic pyroxene, probably diopside.

The granular marble is exposed in an area about 50 by 150 feet a mile and a half south of Red Cliff high on the' slope west of Homestake Creek. A small amount of quartzite is in contact with the marble. The marble is white and fine grained, yet coarse enough to show many calcite cleavage faces. The rock retains traces of the original bedding. Large patches of wollastonite or related mineral are found in the marble. In thin section the calcite grains show the characteristic multiple twinning. Scattered throughout the slide and composing nearly one-fourth of the rock is monoclinic pyroxene with high extinction angle. Qualitative wet tests show this mineral to be iron-bearing diopside. The wollastonitic material has the chemical and physical properties of normal wollastonite. The optical properties in thin section are those of wollastonite, excepting part of the material which shows extinction inclined to traces of the cleavage.

CHAPTER III

IGNEOUS ROCKS

By R. D. CRAWFORD

The large body of granite and other plutonic rocks of the Red Cliff district and nearby areas may represent successive intrusions widely separated in time. Yet all the granite, quartz monzonite, quartz diorite, gneiss, and schist were long exposed to erosion before the deposition of the Sawatch formation of Cambrian age. This is shown by the peneplaned surface of contact between the pre-Cambrian rocks and overlying quartzite. Owing largely to the presence of extensive sediments only small exposures of granite are found within the area mapped.

GRANITE

Though the granite varies considerably in composition, texture, and structure no attempt has been made to map separately the different varieties. They are mainly massive in structure, but locally, and especially near the mouth of Homestake Creek, the granite is strongly gneissoid. Near the border of the intrusion in this locality the granite is interlayered with the older rocks, and it is impossible to place the exact contact between granite and gneiss. The normal granite is mostly of even and medium-sized grain. However, fine-grained and very coarse-grained varieties are found. Though a small amount of pegmatite is present strong dikes of pegmatite having large masses of quartz and feldspar are apparently absent from the Red Cliff district, and aplite dikes are unimportant here.

Biotite, feldspar, and quartz can be megascopically determined in most of the unaltered granite. In a few hand specimens albite twin striations on cleavage faces show part of the feldspar to be plagioclase. Locally, especially near ore veins, kaolin replaces part or all of the feldspar, and a limonite stain is common.

Under the microscope the following minerals can be seen: apatite, zircon, magnetite, pyrite, biotite, muscovite, soda-lime feldspar, alkali feldspar, and quartz, beside several secondary minerals. Three thin sections carry hair-like microlites of an indeterminable mineral that is probably rutile.

Apatite is found in only minute quantity and in small crystals in a few specimens. Zircon is likewise rare, and occurs in minute stout crystals. Magnetite is found in small anhedral grains in about half the specimens; it has probably been derived in part from biotite. Pyrite is very rare.

Biotite, or chlorite that has replaced biotite, is present in nearly every thin section examined. The biotite content is commonly below 10 per cent, though one specimen examined carries about 15 per cent of biotite. Chlorite is a common alteration product, and in some specimens replaces nearly or quite all the biotite. A small amount of muscovite is present in most of the thin sections, and is probably in part a pyrogenetic component.

In some specimens, particularly those from Camp Bryan, the feldspar is nearly all microcline. In others orthoclase occurs in considerable quantity, and the two feldspars may intergrade in the same grain. In a few specimens albite or oligoclase is perthitically intergrown with the potash feldspar. Independent grains of albite have been definitely determined in only one specimen—that from a narrow dike in quartz diorite about 2 miles northwest of Pando. The soda-lime feldspar, which varies from a negligible component in certain varieties of the rock to a quantitatively important component in the alkalicalcic granite that approaches quartz monzonite in composition, is mostly a low-calcium andesine. Kaolin and finely-flaked white mica are common alteration products of the feldspars. A few micrographic intergrowths of quartz and feldspar may be seen, but the quartz is found chiefly in independent grains as the last product of crystallization from the magma. The quartz which commonly shows undulatory extinction, makes up about 15 to 30 per cent of the rock, and will probably average at least 25 per cent.

QUARTZ MONZONITE

In appearance and texture quartz monzonite resembles granite, but it differs from granite in chemical and mineral composition. Typical granite carries alkali feldspar (most commonly potash feldspar) in excess of soda-lime feldspar, while quartz monzonite carries the two kinds of feldspar in nearly equal amounts. Since the identity of most of the feldspar grains and their relative abundance can be determined only with the aid of the microscope quartz monzonite is ordinarily called granite in the field.

In Eagle Canyon, at and near Belden, is a considerable exposure of porphyritic quartz monzonite whose relation to the granite is difficult to determine. A short distance east of Belden the two rocks seem to intergrade, and it is probable that they are different phases of a single intrusion. That both had solidified and suffered erosion long before the sand of the Cambrian quartzite was deposited is shown by the nearly plane contact between the igneous rocks and overlying quartzite. Because of the indistinct boundary between granite and quartz monzonite the latter has not been mapped separately.

The typical quartz monzonite of this area contains reddish feldspar phenocrysts which have a maximum length of about 2 inches, though the average length is probably less than an inch. In places the phenocrystic feldspar composes nearly one-third of the rock mass. Locally the phenocrysts are fewer or absent. Some phenocrysts inclose crystals of biotite and plagioclase, readily seen on cleavage faces. A few phenocrysts are twinned according to the Carlsbad law. The fairly coarse groundmass is composed essentially of biotite, quartz, and bluish feldspar. Under a lens some blue feldspar grains show albite twin striations on cleavage faces. Individual grains of the groundmass are mostly less than a quarter of an inch in diameter. A few small grains of pyrite may be seen in some specimens.

Thin sections contain apatite, zircon, magnetite, pyrite, biotite, muscovite, soda-lime feldspar, potash feldspar, quartz, and several secondary minerals. Apatite is rare in minute prismatic crystals. Zircon occurs sparingly in well shaped crystals. Magnetite is a common accessory in well formed crystals and formless grains. Very few small anhedrons of pyrite are present.

Biotite is the ordinary variety, and forms an average of 10 per cent or less of the rock mass. With the biotite is a small amount of muscovite which is probably in part primary. A small quantity of epidote occurs as an alteration product. The potash feldspar is largely microcline, though many individual grains or crystals are partly microcline and partly orthoclase. With the potash feldspar is a very small quantity of micro-perthitically intergrown albite or oligoclase. The soda-lime feldspar is andesine having medium or low lime content. A common alteration product of this feldspar is a finely flaked mica, probably paragonite. The granular groundmass carries far more soda-lime feldspar than alkali feldspar, but the two kinds are nearly equal in quantity when the rock as a whole—phenocrysts and groundmass—is considered. The quartz content of the rock will run nearly 30 per cent. The quartz shows undulatory extinction and carries numerous liquid and gas inclusions.

West of Eagle River, southwest of the mouth of Rock Creek, the quartz monzonite (here with few phenocrysts) incloses a fine-grained dark-colored rock in masses from a few inches to 5 feet across. An inclusion may show a sharp contact with the coarser rock on one side and grade into the coarser rock on the other side. The fine-grained rock is fissured and faulted and crossed by thin dikes of granite or pegmatite. Thin sections show less quartz, more andesine, and more biotite than the inclosing rock. The inclusions are probably secretions formed from the quartz-monzonite magma in an early stage of consolidation.

QUARTZ DIORITE

Several varieties of quartz diorite are found in the district, but they fall into two general classes: (a) those in which hornblende is an essential component, and (b) those carrying little or no hornblende.

QUARTZ-HORNBLENDE DIORITE

Northwest of Pando this variety of diorite is exposed on both slopes of the valley. Within the valley, throughout a considerable area the same rock is covered by soil and glacial material. Farther west the quartz diorite is overlain by the Cambrian quartzite. The igneous rock was intruded as a stock into the gneiss evidently in pre-Cambrian time. This stock is in turn cut by dikes of pegmatite, aplite, and ordinary red granite. The dikes range from a few inches to several feet in width, and seem to be confined to areas near the border of the stock.

A smaller stock of the same kind of diorite is found on the west side of Eagle Canyon below Red Cliff. Boulders of the same rock may be seen near the southeast rim of Homestake Gulch several miles southwest of Pando.

On fresh surfaces the rock is dark gray, in places almost black. It weathers to lighter and, locally, brownish gray. Though the ratio of light to dark minerals is variable, more than half of some specimens is composed of black hornblende and biotite. These two minerals likewise bear no constant relation to each other; here one, there the other is in excess. The light colored minerals are feldspar and quartz. Under a lens some feldspar cleavage faces show albite twin striations. The quartz diorite is appreciably, though not strikingly, gneissoid in structure.

In thin section zircon, apatite, titanite, magnetite, hornblende, biotite, plagioclase, orthoclase, and quartz are present.

The zircon is partly in minute barrel-shaped crystals, and partly in well shaped prismatic crystals with length five or six times the thickness. Apatite is fairly plentiful, for an accessory component, and in unusually large, though microscopic, crystals. Titanite is rare in formless grains. Magnetite occurs in anhedrons of considerable size in some specimens; a few minute octahedrons can be seen. The apparent practically unaltered character of the hornblende and biotite indicates that the magnetite is mostly an original constituent.

Hornblende is the common green variety, and ranges from about 5 to 50 per cent of the rock volume in specimens examined. The mica is ordinary brown biotite. It varies from a few per cent of the rock volume in specimens having much hornblende to about 20 per cent where the hornblende content is low.

Plagioclase is an important component of all the specimens examined. It is commonly twinned after both albite and pericline laws, and occasional Carlsbad twins are present. Interference colors and extinction angles indicate that the feldspar is andesine. The mineral shows considerable alteration. Orthoclase is present in very small quantity. Quartz varies from 10 or 15 per cent of the rock volume to an almost negligible component.

A small intrusion of somewhat similar rock is found about seven-eighths of a mile southeast of Pando. Common green hornblende composes more than half of the rock mass, and monoclinic pyroxene is an important constituent. The other essential components are orthoclase and plagioclase. The rock is a basic monzonite-diorite which in the proportion of component hornblende lies between the quartz-hornblende diorite and the hornblendite described farther on.

QUARTZ-MICA DIORITE (WITHOUT ESSENTIAL HORNBLENDE)

Beginning several hundred feet west of the Red Cliff depot and extending down Eagle Canyon on both walls, is a quartz-mica diorite almost free from hornblende and texturally and structurally unlike the rock described above. Near the north border of the quartz diorite the rock becomes somewhat monzonitic owing to a considerable content of orthoclase.

The quartz diorite has a dark gray color, massive structure commonly, and medium to very coarse texture. Near the east border of the exposure cleavage faces of black mica throughout the rock reach a diameter of nearly half and inch; at the same place are segregations of almost pure biotite whose leaves are nearly an inch across. Here also are small patches of pyrite that probably crystallized from the molten rock magma. Near this border also narrow dikes of aplite and pegmatite cut the quartz diorite.

Throughout the exposure biotite is an important constituent of the rock. It ranges perhaps from 15 to 30 per cent of the rock volume and averages probably about 25 per cent. Aside from biotite and the small amount of pyrite mentioned, bluish-gray quartz and feldspar are the only minerals identifiable in hand specimens. Some feldspar cleavage faces show albite twin striations. Individual grains of quartz and feldspar are commonly less than a quarter of an inch in diameter.

Under the microscope may be seen zircon, apatite, magnetite, biotite, feldspar, and quartz; a few small grains of hornblende were noted in only one specimen. Zircon and apatite are very rare. Magnetite is present in every thin section examined, and for an accessory component is plentiful. It occurs in both euhedrons and anhedrons, and may be in part secondary.

Biotite is the ordinary variety, brown in thin section. In some sections it shows alteration to epidote or muscovite. Part of the magnetite observed is probably an alteration product from biotite.

In nearly every thin section examined the feldspar is exclusively plagioclase. Not infrequently pericline or Carlsbad twinning is combined with the universal albite twinning. A few determinations by the Michel-Levy method show the feldspar to be andesine. It is mostly a highly calcic variety having a composition of approximately Ab_1An_1, though in some specimens the andesine has a higher soda content. This feldspar shows common alteration to a finely flaked mica, probably paragonite. Epidote, calcite, and kaolin are less common alteration products. One specimen from near the northwest border of quartz diorite is quartz-monzonitic in composition, having potash feldspar only slightly subordinate to the andesine. In this specimen both orthoclase and microcline are present.

Quartz is abundant and composes about 25 to 30 per cent of most of the specimens examined. It commonly shows slight to moderate undulatory extinction. In some specimens the quartz is intergrown with andesine.

Gneissoid quartz-mica diorite.—A small intrusion of quartz diorite north of Petersons Gulch megascopically resembles the hornblende-bearing quartz diorite, but no hornblende is seen in the one thin section examined. It also carries a high content of quartz and biotite. The rock is somewhat porphyritic in texture, but probably should be included with the pre-Cambrian granular intrusive rocks.

SYENITE

Less than a fourth of a mile southwest of the dense marble mentioned in the preceding chapter is a dike of basic syenite. The rock is of dark color and very fine grain. Only biotite and reddish-gray feldspar can be identified in the hand specimen.

Nearly half the volume of a thin section is orthoclase which incloses numerous small apatite crystals. A very little plagioclase is present. Biotite and hornblende are abundant in nearly equal quantity. The rock carries much epidote.

HORNBLENDITE

About 2 miles N. 70° W. of Pando, on the east side of Homestake Gulch, is a small exposure of hornblendite. Probably 80 per cent or more of the rock is hornblende. Pyrite, light brown mica, and andesine are the other components. A short dike of similar hornblendite having a larger proportion of hornblende is found about 2.5 miles northwest of Pando, east of the highway. Near the last mentioned occurrence is a small irregularly shaped intrusion of hornblendite in the gneiss. It is cut pegmatite dikes, and incloses a few quartz lenses 2 or 3 inches in diameter. Greenish-black hornblende, in grains one-half to three-fourths inch in diameter, forms the bulk of this variety of hornblendite. A noticeable feature, is the large number of small biotite flakes that cut across the hornblende cleavage planes at a high angle. Other small flakes, singly and in aggregates roughly parallel to the hornblende cleavage faces, are inclosed by hornblende. A third occurrence of biotite shows this mineral in large grains intergrown with the hornblende in a manner that gives a mottled luster to the mica cleavage faces. The microscope discloses small quantities of titanite, magnetite, feldspar, and quartz.

PERIDOTITE

Near the last named locality, east of the highway and in line with Homestake Gulch, are many boulders of a peculiar peridotite. They are evidently boulders of weathering that have traveled little if at all; similar boulders have not been found elsewhere by the Survey party.

The peridotite is a heavy, tough, grayish-black rock with striking mottled luster. The peculiar luster is best shown by turning a freshly broken surface toward the sun, when it is seen that nearly half the surface is covered by patches of reddish-brown mica whose cleavage faces reflect the light. The mica grains will average an inch in diameter while individual patches are less than a sixteenth of an inch across. The mica is intergrown with or incloses other minerals that interrupt the reflection. This, taken with the different orientation of different mica grains, explains a large part of the peculiar luster. In the same rock large pyroxene grains with their cleavage faces interrupted by intergrown minerals show similar mottling and duller luster.

Magnetite, olivine, pyroxene, phlogopite, and serpentine are seen under the microscope. The magnetite is fairly plentiful in small grains; it may be in part secondary. Olivine perhaps originally exceeded any other component and is still abundant; it is in part replaced by serpentine. Part of the pyroxene is monoclinic, but some grains with parallel extinction may be an orthorhombic variety. The phlogopite is reddish-brown and different from ordinary biotite; it reacts strongly for fluorine. Both phlogopite and pyroxene poikilitically inclose olivine and serpentine.

PORPHYRY AND FELSITE

In this report the names porphyry and felsite are used for dense igneous rocks as textural terms without reference to mineral composition. **Porphyry** is accordingly a dense rock in which part of the material of the original magma solidified as crystals large enough to be seen by the naked eye. These crystals, called phenocrysts, may in their aggregate form either a small fraction or the greater part of the rock mass. The rest of the rock, called groundmass, is dense or microcrystalline. **Felsite** is a light colored rock composed essentially of dense material like the groundmass of porphyry. It contains few or no phenocrysts. Obviously porphyry and felsite may grade into each other.

In many parts of central Colorado dikes or sheets of porphyry are common. They are of comparatively late age, and some of them were almost certainly intruded in Tertiary time. In the Red Cliff district the porphyries were intruded into the youngest Paleozoic sediments, but there are no relationships by which the exact age of the intrusions can be determined.

In this district exposed surfaces of porphyry and felsite are relatively small. For the most part these rocks occur as intrusive sheets seen only on their outcropping edges. In the vicinity of Gilman and Red Cliff a porphyry sheet is found a few feet above the top of the Leadville limestone where it was intruded nearly parallel to the bedding of the inclosing rocks. This sheet is exposed at intervals through a distance of about 10 miles; it is evidently continuous below the surface from the southeast part of the mapped area to a point a mile or more north of Gilman. The thickness varies: in the Wilkesbarre shaft of the Eagle mines, at Gilman, the sheet is said to be about 55 feet thick; in places it is much thicker or thinner. Locally the sheet passes from one horizon to another only a few feet higher or lower, but it is generally underlain by black shale. In the Liberty incline, near Red Cliff, the intrusion is split by about 5 feet of shale. A mile or two southeast of Red Cliff this prophyry forms three sheets for a short distance instead of one.

South of Yoder Gulch and north of Pando are a few outcrops of a porphyry sheet that was intruded into or just below the Sawatch quartzite. Near the southeast part of the mapped area other sheets occur at higher horizons than those mentioned. Very few dikes in the pre-Cambrian rocks and crosscutting the sediments have been found. The higher sheets can be traced to the outcrops of porphyry of the Tenmile district. Alluvium and landslides near the southeast corner of the area mapped conceal the Gilman sheet and others in the lower part of the Weber beds, but these sheets are probably connected with the laccolith of Chicago Mountain. The center of this laccolith is about 5 miles southeast of Pando and 4 miles west of Robinson. The Tenmile district map published by the United States Geological Survey shows the laccolith to be at least 2,000 feet thick where it is deeply eroded by Eagle River.

Excepting the quartz-monzonite porphyry and quartz-diorite porphyry described in later pages all the porphyry of this district has a dense groundmass in excess of the volume of phenocrysts. Owing to fineness of grain in the groundmass and the generally altered character of the entire rock of most exposures it is difficult or impossible to name the rock exactly. The Gilman sheet and a few others, as well as one or two dikes, probably have orthoclase greatly in excess of soda-lime feldspar, and hence are rhyolite porphyry or rhyolite. Some of the sheets seem to carry nearly equal amounts of orthoclase and soda-lime feldspar. These relationships, together with the presence of quartz and the character and proportion of groundmass, make the name quartz-latite porphyry, or dellenite porphyry, appropriate.

The least altered specimens were collected near the southeast corner of the area mapped, where the rock does not differ greatly from the quartz-monzonite porphyry described below. The reddish groundmass of these specimens is dense, and forms more than one-half the rock mass. The phenocrysts are of biotite, quartz, and white feldspar, in part plagioclase. The quartz and feldspar phenocrysts have a maximum diameter of three-eighths of an inch. Under the microscope the groundmass is seen to contain quartz, orthoclase, and plagioclase. Secondary limonite, magnetite, and sericite are present.

Throughout the rest of the district the porphyry is greatly altered, apparently in large part by warm waters. The biotite has been changed principally to chlorite and muscovite, while the feldspar of phenocrysts and groundmass shows much alteration to kaolin and white mica. A thin section from a drill core that came from the sheet below the Liberty tunnel at a distance of more than 1,000 feet from the outcrop, shows about the same degree of alteration as the freshest specimen collected at the outcrop of the same sheet north of Gilman. Beside the minerals mentioned secondary calcite and magnetite are often seen. Most of the porphyry has a greater proportion of groundmass than has the variety in the southeastern part of the district, and the phenocrysts are smaller.

In thin sections of the porphyry are seen apatite, magnetite, plagioclase, orthoclase, and quartz, beside several secondary minerals. Apatite is very rare and in minute crystals. Though the magnetite may be partly primary the most of it has probably been formed by the alteration of biotite. In many specimens limonite is a common secondary mineral.

The feldspar phenocrysts are nearly all plagioclase, probably andesine. This feldspar is commonly much altered, and shows calcite, kaolin, and a finely flaked white mica, probably paragonite. The biotite is the ordinary brown variety, and rarely appears in fresh condition. Chlorite and muscovite, with a little magnetite, commonly occur with the biotite or completely replace it as pseudomorphs. The quartz phenocrysts are generally rounded and embayed by resorption.

A common type of groundmass is micropoikilitic. Microlites of plagioclase with roughly rectangular outline are abundant as inclusions in microscopic grains of orthoclase. In many instances single small grains of quartz inclose feldspar microlites and add to the micropoikilitic effect. The microlites show inclined extinction and have a very low birefringence; they are hence probably andesine. Though individually small, they probably form one-third of the total feldspar of the groundmass. Adding this to the phenocrystic plagioclase it is probable that the combined amount of plagioclase in phenocrysts and groundmass nearly equals the quantity of orthoclase. If these estimates are correct the rock of this type is a dellenite porphyry, or quartz-latite porphyry. A second type of groundmass is micrographic. Individual grains are quartz and orthoclase and, in some instances, plagioclase. Some of the rock with this texture, because of its high plagioclase content, should be classed with the quartz latites; some, especially the non-porphyritic felsite, carries very little plagioclase, and is properly called rhyolite. White mica, kaolin, iron oxide, and calcite are common secondary minerals in the groundmass.

QUARTZ-MONZONITE PORPHYRY

In the southeast part of the area mapped are several intrusive sheets of porphyries that carry a larger proportion of phenocrysts than the rocks described above. About one-half to one-third of the rock is groundmass, mostly coarse microcrystalline. Locally the groundmass exceeds the volume of phenocrysts, and the rock becomes dellenite porphyry. Though several varieties are present they are all characterized by quartz and nearly equal quantities of alkali feldspar and soda-lime feldspar, and they carry many crystals of black minerals. These are features of quartz-monzonite porphyry. These intrusions are continuous with the sheets and the Chicago Mountain laccolith of the Tenmile district where they were mapped, in 1881-2, by the United States Geological Survey. The rocks were described by Emmons[2] under the name "diorite porphyries" in accordance with the customary classification of the time. Chemical analyses[3], as well as mineral determinations, indicate quartz monzonite porphyry.

Quartz phenocrysts are abundant; many of them show resorption. In the "Lincoln porphyry" part of the phenocrystic feldspar is orthoclase, but most of the feldspar phenocrysts of all varieties are plagioclase, evidently andesine. Biotite, in small crystals, is common. Accessory minerals noted in thin section are allanite, apatite and magnetite. The microgranular groundmass is composed essentially of quartz and orthoclase, with considerable plagioclase in some specimens. Calcite, chlorite, kaolin, epidote, and finely flaked white mica are common secondary minerals.

QUARTZ-DIORITE PORPHYRY

One small exposure of quartz-diorite porphyry is found about two and a half miles southeast of Pando. The intrusion cuts across the bedded rocks and is either a dike or a stock. The rock differs from the quartz-monzonite porphyry described in having a higher proportion of plagioclase, there being about three times as much plagioclase as orthoclase. About three-fourths of the rock is composed of phenocrysts of quartz, plagioclase, biotite, and hornblende. The quantity of hornblende is almost negligible. Optical properties of the plagioclase indicate andesine. The coarse-microgranular groundmass is composed mostly of quartz and orthoclase. Secondary minerals seen in thin sections are chlorite, epidote, magnetite, calcite, and kaolin.

---

[2]Emmons, S. F., Tenmile district special folio (No. 48), U. S. Geol. Survey 1898.
[3]See U. S. Geol. Survey Bull. 591, p. 129, B and C, 1915.

CHAPTER IV

PALEOZOIC SEDIMENTARY ROCKS[*]

By Russell Gibson

The sedimentary rocks of the Red Cliff district include Cambrian, Ordovician (?), Devonian, Mississippian, and Pennsylvanian strata. The total thickness of these beds within the area mapped is about 4,600 feet. Except where streams have cut deep enough to reveal the underlying pre-Cambrian igneous and metamorphic rocks, most of the surface exposures are of Paleozoic sediments. Surficial material, such as glacial debris, alluvium, and landslides, has in places obscured all the older rocks. A scarcity of fossils has made age determinations on faunal evidence exceedingly difficult. The correlations are supported partly by lithologic similarity and partly by stratigraphic position.

CAMBRIAN AND ORDOVICIAN OR DEVONIAN SEDIMENTS

SAWATCH QUARTZITE

DESCRIPTION

The sediments herein referred to as the Sawatch quartzite are about 356 feet thick, and consist of the Paleozoic sedimentary rocks older than the Leadville limestone. The Sawatch lies unconformably on the pre-Cambrian igneous and metamorphic rocks.

At the base of the Sawatch, a conglomerate about 9 feet in thickness is made up chiefly of rounded quartz grains whose maximum diameter is one-half inch. Most of the material is finer than this, however, and contains fragments of feldspar. The cementing material is silica and kaolin. The conglomerate grades upward into a fine-grained, light-colored quartzite of which most of the superjacent 210 feet is composed. A few beds are less firmly cemented than typical quartzite; a few thin sandy shale beds are interlarded; and one brownish sandy limestone bed occurs 76 feet above the granite, in a section measured in Eagle Bird Gulch near Gilman. The beds which are not well cemented are commonly ferruginous and darker in color. Approximately 220 feet above the granite there is a marked change in the character of the sediments. For the next 105 feet the rocks are dominantly thin-bedded impure limestones, sandstones, and shales. The color varies from gray to brownish-red; some of the beds are mottled reddish-brown and green. The best defined beds are gray to light-brown or brownish-red magnesian limestones, most of which are siliceous, and some are ferruginous as well. At the top of the section there are 30 to 35 feet of white to gray quartzite, the upper portion of which is calcareous and conglomeratic with pebbles up to three-eighths of an inch in diameter. Except for the calcareous cement in the extreme upper part of this 30 feet, the beds are very similar to the quartzite at the base of the series. The following section is exposed in Eagle Bird Gulch:

---

[*]In this chapter the writer has quoted freely from an unpublished thesis by E. A. Hall, a portion of which is devoted to the stratigraphy of the Red Cliff district. This thesis was submitted to the faculty of the Graduate School of the University of Colorado in partial fulfillment of the requirements for the degree Master of Science. Professor W. C. Toepelman, of the University of Colorado, identified the fossils named in this chapter.

Section of Sawatch Quartzite at Eagle Bird Gulch

Feet

(Leadville limestone uncomformable upon)

29. Quartzite, varying from white through gray to brown, coarse grained, containing near the top pebbles up to three-eighths inch in diameter. Upper part calcareous_____ 14.5
28. Quartzite, white, hard, fine grained, well bedded_____ 15.7
27. Covered by talus_____ 24.0
26. Shale, mottled reddish brown and green_____ 6.6
25. Quartzite, dense_____ 0.7
24. Shale, mottled reddish brown and green_____ 0.7
23. Sandstone and sandy shale, green, dark purple, and gray; sandstone, coarse and not well cemented_____ 8.0
22. Quartzite, dark purple, hard, fine grained, with a few, thin shale beds; bedding somewhat wavy; weathers brown or dark purple__ 3.4
21. Sandstone, dense, even grained, argillaceous, calcareous, magnesian; in alternate brown and dark red, thin laminae_____ 0.6
20. Limestone, dolomitic, ferruginous, sandy, dark red, hard, thick bedded with a few, thin shale beds; bedding somewhat wavy; weathers brown and dark purple_____ 19.0
19. Limestone, gray, crystalline; sandstone, dark red, calcerous, magnesian; shale, gray, sandy, calcareous, laminated; beds wavy and uneven in places_____ 5.2
18. Limestone, shale, and sandstone, similar to No. 17_____ 2.5
17. Limestone, crystalline, sandy, gray with brown mottling; shale, laminated, friable, arenaceous, calcareous; sandstone, thin bedded, greenish gray; contains *Saukia* near top_____ 23.4
16. Limestone, gray to light brown, dolomitic, ferruginous, slightly arenaceous; shale, gray, slightly calcareous, arenaceous, laminated 5.5
15. Sandstone, gray, calcareous, laminated_____ 4.2
14. Quartzite, brownish gray, containing streaks less firmly cemented; interstratified with purplish arenaceous shale_____ 30.1
13. Quartzite, white, dense_____ 8.6
12. Sandstone, firmly cemented, medium grained, interstratified with white fine-grained quartzite; well jointed. Some brecciated zones recemented with silica_____ 9.2
11. Quartzite, white, fine grained, well jointed in beds 1 to 15 inches thick _____ 24.1
10. Quartzite, porous, mineralized_____ 6.3
9. Quartzite, very dense, firmly cemented, in beds one-fourth inch to 40 inches thick, interstratified with a few beds of porous, heavily iron-stained quartzite_____ 49.6
8. Sandstone, friable, iron-stained; grades laterally into quartzite____ 4.1
7. Quartzite, very dense, firmly cemented; contains an 18-inch bed of brownish arenaceous limestone_____ 16.9
6. Quartzite, loosely cemented, iron-stained; micaceous in places; less resistant to erosion than adjacent beds_____ 9.5
5. Quartzite, brownish, dense, interstratified with thin beds of arenaceous shale; micaceous bed 3 to 5 inches thick near top_____ 22.7
4. Quartzite, medium grained, tightly cemented_____ 17.7
3. Conglomerate, arenaceous, easily eroded; arenaceous shale at base _____ 3.2
2. Quartzite, light gray, coarse grained, firmly cemented, interstratified with thin beds of conglomerate_____ 10.7
1. Conglomerate, gray to brown, fairly well cemented, containing pebbles up to one-half inch_____ 8.7

356.0

(Unconformable upon pre-Cambrian igneous and metamorphic rocks.)

AGE

Cambrian.—About 252 feet above the pre-Cambrian rocks, fragments of *Saukia pepinensis* and a few brachiopods, *Obolidae* and *Westonia ella*, were found. For nearly 50 feet above this point the beds are similar to those from which these fossils were collected, and careful examination revealed no unconformity. The quartzite beds in the lower part of this section are lithologically identical with the Sawatch as described in neighboring areas[5]. The beds in which, and just above which, the fossils were found are very similar to the upper part of the Sawatch in most of these same areas. Therefore, in view of the stratigraphic position of this division, the fossil evidence, and the lithologic character of the beds, the age of the lower part of the series, that is, the lower 302 feet, is provisionally regarded as Sawatch or upper Cambrian.

Ordovician or Devonian.—Although it is reasonable to suppose that the first 302 feet of this formation in the section measured at Eagle Bird Gulch is upper Cambrian, it is entirely possible that much of the remaining 54 feet may be either Ordovician or Devonian in age. It is certain that all or a portion of the top 30 feet, which is quartzite, is younger than the Cambrian, because a few corals which appear to be *Zaphrentis* sp. were found 2 feet below the top. Moreover, the 30 feet of quartzite at the top of the debatable beds is very different from the limestones, sandstones, and shales where the Cambrian fossils are found. Immediately below this top quartzite member the beds are, in most places, obscured by soil; hence no evidence of an unconformity—which may be present—was found. Data are insufficient to correlate these 30 to 35 feet of quartzite with the Ordovician or Devonian "Parting quartzite" of other districts. It is certain only that this quartzite is post-Cambrian but not younger than Devonian. The contact between the post-Cambrian and the Cambrian sediments is probably at the base of this quartzite member or a few feet below it. On the geologic map all the sediments below the Leadville limestone are included in the term "Sawatch quartzite."

DEVONIAN-MISSISSIPPIAN SEDIMENTS

LEADVILLE LIMESTONE

DESCRIPTION

The dark-gray, dense, thick-bedded limestone which immediately overlies the coarse-grained quartzite beds, the upper few feet of which are calcareous, conglomeratic, and not well cemented, is considered to be the base of the Leadville limestone. The upper limit of the Leadville is taken at the bottom of the group of alternate thin-bedded black shales and dark-gray, thin-bedded, dense limestones. The total thickness of the Leadville

---

[5]Emmons, S. F., Geology and mining industry of Leadville, Colorado: U. S. Geol. Survey Mon. 12, 1886, p. 58.
Eldridge, G. H., U. S. Geol. Survey Geol. Atlas, Anthracite and Crested Butte folio (No. 9), 1894.
Emmons, S. F., U. S. Geol. Survey Geol. Atlas, Tenmile folio (No. 48), 1898.
Spurr, J. E., Geology of the Aspen mining district, Colorado: U. S. Geol. Survey Mon. 31, 1898, p. 4.
Girty, G. H., Carboniferous formations and faunas of Colorado: U. S. Geol. Survey Prof. Paper 16, 1903, p. 140.
Worcester, P. G., Geology and ore deposits of the Gold Brick district, Colorado: Colo. Geol. Survey Bulletin 10, 1916, pp. 51-52.

measured at Eagle Bird Gulch is 199.5 feet; at Rock Creek, 222 feet; and at a point about one mile south of the mouth of Rule Creek, at least 170 feet.

At Eagle Bird Gulch and at Rock Creek the porphyry is found intruded at, or very close to, the top of the formation here considered. As mentioned under the description of the Sawatch, there is an evident disconformity at the base of the Leadville. The lower 70 to 75 feet of the formation is made up of even-bedded, dense, fairly hard, dark-gray, crystalline, magnesian limestone or dolomite, which weathers light gray. Some of the thin beds that are well jointed weather into small blocks. The weathered surfaces of other beds exhibit a somewhat knotty or lumpy form. A few narrow chert layers are present and also a few thin beds of arenaceous shale. Above this 70 to 75 feet of limestone or dolomite is a remarkably persistent zone of quartzite and breccia, seemingly marking an unconformity. Five good exposures of this quartzite were seen between Rock Creek and a point 2 miles south of Red Cliff. The quartzite was found 93 feet above the base of the Leadville at a point one mile south of the mouth of Rule Creek. The lower part of the zone is a quartzite bed from 10 to 28 inches thick, composed of fine to coarse well rounded quartz grains firmly cemented with both siliceous and calcareous material. The color is gray to brown. The upper part of the zone is breccia from 5 to 12 feet thick composed of quartz grains and angular, rounded, and flattened fragments of limestone and chert bound together by a cement that is calcareous in some places and siliceous in others. Some of the large fragments of limestone are 5 inches long, but the average length of the fragments is less than an inch.

. The upper part of the Leadville, 110 to 145 feet in thickness, differs from the lower part in several respects. It is massively bedded; more coarsely crystalline; contains more vugs; and has an irregularly laminated structure, commonly parallel to the bedding, which appears on the weathered face of the rock as closely spaced alternate dark and light lines. In places there are four or five white lines to an inch; elsewhere they are farther apart. Not uncommonly the streaks become very irregular, and the limestone appears spotted rather than banded. The mineral composition of the light and dark material is identical. The miners term this rock "zebra" limestone. The vugs are irregular in size, rarely 4 inches long, and are lined with dolomite and quartz crystals. A few thin chert layers are in the middle of the upper member of the formation.

The entire Leadville is dolomitic and siliceous; traces of aluminum and iron are present in most of the beds; and a considerable amount of phosphate was found in a specimen from a thin bed of limestone 6 to 8 feet above the breccia. No outcrop of the Leadville limestone of the characteristic type has been found east of the river between Silver Creek and the small exposure of this limestone a mile east of the mouth of Rule Creek. Through much of this distance (about four miles) the bedrock is covered by landslides, moraines, and other surficial material. Wherever the bedrock is exposed in this part of the district the limestone has been replaced by jasperoid quartz. The character and extent of the replacement are described in Chapter VI. Below is a section of the Leadville limestone measured at Eagle Bird Gulch:

Section of Leadville at Eagle Bird Gulch

Feet

(Pennsylvanian unconformable upon.)

21. (Porphyry 80 feet thick)
20. Limestone (similar to bed No. 16 below)_____ 44.4
19. Limestone, gray, medium- to thick-bedded, "zebra" texture; surface furrowed by weathering_____ 19.0
18. Limestone, well bedded; joints prominent_____ 2.5
17. Limestone, dark gray, cherty_____ 3.0
16. Limestone, gray to grayish black, crystalline, containing vugs up to 4 inches in width lined chiefly with quartz, calcite, and dolomite; bedding indistinct. Limestone grades laterally into coarsely crystalline, finely banded, "zebra" limestone; surface furrowed by weathering_____ 37.5
15. Limestone, dark gray; weathers white to yellow; contains black chert _____ 2.6
14. Breccia, dark gray, made up of fragments up to 5 inches long of limestone, chert, and quartzite; light-colored, medium-grained quartzite; and gray shale_____ 13.5
13. Quartzite, gray, coarse-grained_____ 1.8
12. Limestone, gray, arenaceous; has contorted appearance_____ 1.0
11. Limestone, light- to dark-gray, even-bedded; weathered surface exhibits alternate light and dark beds_____ 9.7
10. Limestone, grayish black, dense, even-bedded; contains vertical veins of calcite and dolomite_____ 7.5
9. Limestone, grayish black, cherty, even-bedded; has contorted appearance; upper part weathers into blocks_____ 7.3
8. Limestone, brown, arenaceous, argillaceous_____ 1.3
7. Limestone, grayish black, dense; has a hard, chert-like appearance; contains veins of calcite and dolomite, and vugs with crystals of calcite and dolomite_____ 4.7
6. Limestone, dark gray, containing white to yellowish veins of carbonate. Some beds have a knotty appearance; others are smooth, darker in color, and weather into blocks_____ 9.4
5. Alternate beds of dark-gray, dense, arenaceous limestone, and arenaceous, thinly laminated, greenish shale_____ 10.0
4. Limestone, dark gray, well bedded; has a knotty, wavy appearance; weathers in slabs with rounded edges; contains a few narrow chert layers _____ 13.7
3. Shale, gray to light brown, calcareous, arenaceous, unevenly bedded _____ 0.7
2. Limestone, gray, dense, fine-textured; weathers into blocks_____ 6.0
1. Limestone, brownish-gray, dense, unevenly bedded_____ 3.9
                                                              _____
                                                              199.5

(Unconformable upon Sawatch.)

AGE

The following fossils were found in the basal part of the Leadville:
Spirifer whitneyi, var. animasensis Girty
Spirifer sp., may be Sp. coniculus Girty
Schuchertella sp., resembles Sch. chemungensis Conrad
Crania sp.
Fenestella   ?
Pelecypods, two indeterminate species
Fragments of crinoid stems
Zaphrentis   ?

Two specimens of *Spirifer centronatus* Winchell were given to the Survey party by Dr. I. A. Ettlinger, who found them a few feet above the breccia.

A few corals, probably *Zaphrentis* sp., and one brachiopod, which may be *Chonetes illinoisensis* Worthen, were found near the top of the Leadville.

Fossil species collected are too few to justify a correlation of these beds with the Leadville limestone in other regions with any degree of certainty on faunal evidence alone; but there are other factors which will support a tentative correlation. The upper part of the formation at Red Cliff is lithologically similar to the Blue, or ore-bearing, limestone of the Leadville district, which Emmons[6] has determined as Lower Carboniferous (Mississippian). Girty[7] has accepted Emmons' correlation. Emmons[8] redescribed this formation under the title "Blue or Leadville limestone" in his work on the Tenmile district. Moreover, as is pointed out below, under the discussion of the Pennsylvanian sediments, the Leadville limestone at Red Cliff is overlain by beds which are strikingly similar to those which overlie the Blue limestone of the Leadville district and the Blue or Leadville limestone of the Tenmile district. In view of the fossil evidence found in the upper part of the Leadville, the lithologic similarity to other Mississippian formations in nearby districts, and its stratigraphic position, that part of the Leadville limestone above the breccia zone or unconformity is regarded in this report as Mississippian.

The lower part of the Leadville is tentatively placed in the Devonian for two reasons. An unconformity 70 to 75 feet above the base separates the lower from the upper part; and the only diagnostic fossils identified with certainty, found below the unconformity, are Devonian brachiopods, *Spirifer whitneyi*. Girty[9] mentions that a distinctive Devonian fauna occupies the lower part of the Leadville limestone of central Colorado and the lower part of the Ouray limestone, which is the equivalent of the Leadville. It is also noteworthy that Girty believes that deposition was not quite continuous when the Leadville and the Ouray formations were laid down, and that a slight unconformity may exist between the strata characterized respectively by the Devonian and Mississippian faunas in the midst of the Ouray and Leadville formations. There is little doubt that an unconformity exists at Red Cliff 70 to 75 feet above the base of the Leadville formation.

PENNSYLVANIAN SEDIMENTS

DESCRIPTION

The thin beds of black shale and dark-colored limestone overlying the massive light-colored upper beds of the Leadville formation are taken as the base of the Pennsylvanian. In most places in the district the porphyry sill is found at, or a little above, the contact between the Mississippian and Pennsylvanian. Most of the following description is based on a section measured at Rock Creek.

For approximately 175 feet above its base the Pennsylvanian is chiefly thinly laminated black carbonaceous shale interlarded with four- to twelve-inch beds of dense, dark-gray to black limestone. The shale contains a

[6]Emmons, S. F., Geology and mining industry of Leadville, Colorado: U. S. Geol. Survey Mon. 12, 1886, p. 66.
[7]Girty, G. H., Carboniferous formations and faunas of Colorado: U. S. Geol. Survey Prof. Paper 16, 1903, pp. 163, 222.
[8]Emmons, S. F., U. S. Geol. Survey Atlas, Tenmile folio (No. 48), 1898.
[9]Girty, G. H., Carboniferous formations and faunas of Colorado: U. S. Geol. Survey Prof. Paper 16, 1903, pp. 162-163.

little disseminated pyrite. A few fossils were found. In this section there are a very few dark-colored quartzite beds 1 to 2 feet thick, and near the base a few unimportant thin layers of impure coal. About 175 feet above the base the character of the sediments changes. The limestone beds become ,fewer and disappear; the shales are lighter in color and become arenaceous and micaceous; gray to greenish micaceous sandstone and quartzite beds appear, some of which, higher up, become conglomeratic. At 275 feet the dominant sediments are quartzites and conglomerates. Beginning between 300 and 400 feet above the bottom of the Pennsylvanian

Figure 8. Pennsylvanian sediments, looking north from Rex
Shows shale alternating with sandstone and conglomerate

and extending upward about 450 feet, the sediments are dark or reddish-brown, in contrast with the light-colored beds below and above. The rocks are mainly conglomerates and sandstones and are more or less ferruginous; shales are subordinate. (See Fig. 8) Pebbles in the conglomerate are chiefly quartz, varying in size up to 2 inches in diameter. Fragments of partly altered orthoclase and plagioclase are common. Muscovite, much of it secondary, is abundant in all of the rocks. The interstices between grains in both sandstones and conglomerates contain sericite.

The color of the sediments changes again, about 900 feet above the base, to greenish-gray and gray. For the next 1,600 feet the Pennsylvanian is composed essentially of shale, conglomerate, sandstone, limestone, and quartzite, decreasing in abundance in the order named. Cross-bedding is displayed in places, and many of the beds are lenticular. The shale is micaceous and arenaceous, and grades into sandstone. Sandstone and conglomerate of many degrees of coarseness are found. Pebbles up to 3 or 4 inches in diameter are not uncommon; the maximum size of boulders is 15 inches. The pebbles and boulders are for the most part quartz, quartzite, granite, and gneiss. Feldspar is a common constituent of both sandstone and conglomerate. Pink feldspar is sufficiently abundant in some beds to give a pink tinge to the rocks.

Lenticular dolomitic limestones are prominent but irregularly developed in the upper part of the Pennsylvanian. They are 6 to 50 feet thick; and, since they are rather resistant to weathering and erosion, they stand out in cliffs up to 40 feet high and make one of the most conspicuous features of the topography. The beds are dark to light gray in color, and in places contain chert. In this upper part most of the sediments not limestones are coarse conglomerates and sandstones similar to those a little lower down in the section.

The total thickness of the Pennsylvanian exposed in this district is about 4,000 feet.

AGE

The following fossils were collected from the lower 3,500 to 3,700 feet of the Pennsylvanian. These sediments probably include the Weber shales and Weber grits of the Tenmile district[10]:

Orbiculoidea (cf. manhattanensis M. and H)
Lingula carbonaria Shumard
Orbiculoidea convexa Shumard
Derbya crassa ? (M. and H.)
Meekella striaticostata Cox
Productus cora D'Orbigny
Productus punctatus Morton
Productus inflatus McChesney
Marginifera ingrata Girty
Chonetes geinitzianus Waagen
Reticularia perplexa McChesney
Spirifer cameratus Morton
Spirifer boonensis ? Swallow
Spiriferina sp. (may be Spiriferina kentuckyensis)
Rhombopora lepidodendroides? Meek
Bryozoa undetermined
Allerisma terminale Hall
Nucula ventricosa Hall
Nucula sp. (cf. N. parva McChesney)
Leda bellistriata ? Stevens
Pleurophorus (cf. P. subcostatus M. and W.)
Myalina cuneiformis Gurley
Aviculopecten rectilaterarius Cox
Aviculopecten occidentalis ? Shumard
Zaphrentis undetermined
Campophyllum ?
Chaetetes milleporaceous ? Milne
Monoilopora (cf. prosseri Beede)
Crinoid stems (several types)

[10]Emmons, S. F., U. S. Geol. Survey Geol. Atlas, Tenmile folio (No. 48), 1898.

The uppermost beds of the **Pennsylvanian** system exposed in the Red Cliff district include a portion of **the Maroon** of the Tenmile district. From these sediments the following fossils were collected within the area mapped for this report:

Productus cora D'Orbigny
Meekella striaticostata Cox
Composita subtilita (Hall)
Reticularia perplexa McChesney
Fusulina cylindrica F. de Waldheim
Phillipsia sp.?
Gastropoda undetermined

The fossils in these two lists are among those which have been found in the Pennsylvanian in the Tenmile and Leadville districts[11]. The belief that this formation is Pennsylvanian finds support also in the fact that, in the eastern part of the Red Cliff district, the upper beds are continuous with the Weber and Maroon beds of the Tenmile district which Emmons has determined as Upper Carboniferous, or Pennsylvanian. For several reasons no attempt has been made to divide the Pennsylvanian into Weber shales, Weber grits, and Maroon, as Emmons has done in the Tenmile and Leadville districts. In the Red Cliff district a division of the lower part of the Pennsylvanian into "shales" and "grits" could not be established because the change from carbonaceous shale and limestone below to arenaceous shale, sandstone, and conglomerate above was a gradual one. Fossils found in the Weber and Maroon appear to belong to the same fauna, and furnish no basis for a division between these two formations. Study of the limestone members of the upper part of the Weber and of the Maroon has been insufficient to confirm Emmons' conclusion[13] that the limestones furnish the safest means of distinguishing one formation from the other.

---

[11]U. S. Geol. Survey Prof. Paper 16, pp. 242-243.
[12]U. S. Geol. Survey Geol. Atlas, Tenmile folio (No. 48).

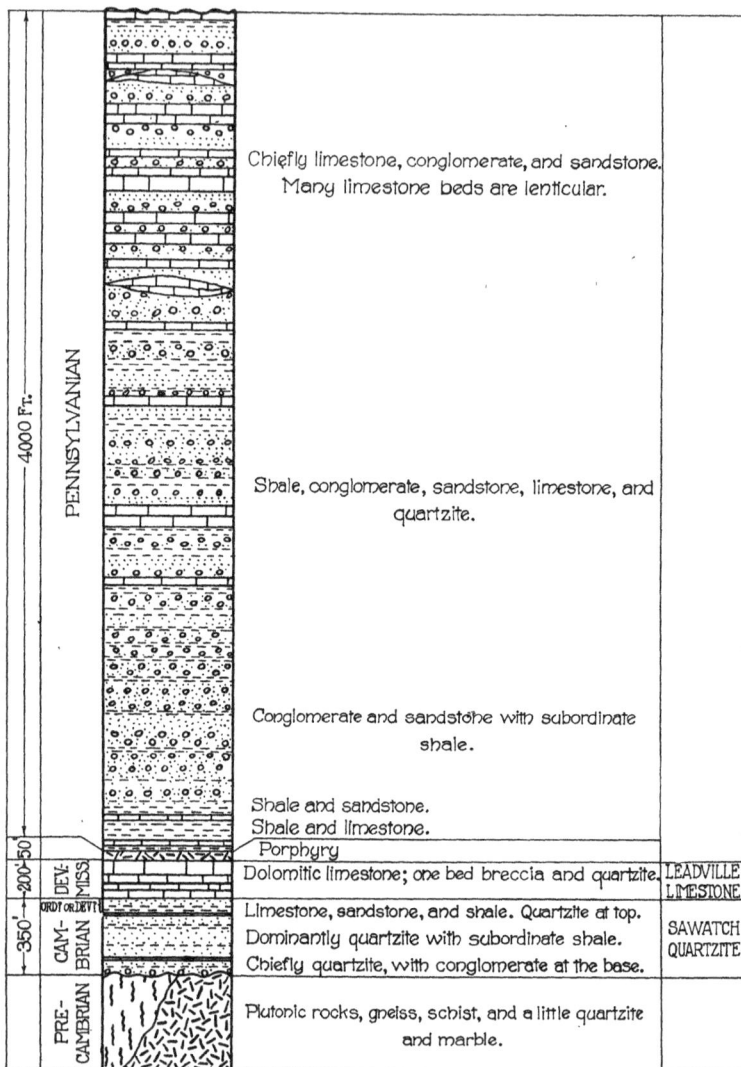

Figure 9. Columnar section of the Paleozoic sediments in the Red Cliff district

CHAPTER V
STRUCTURAL GEOLOGY
By R. D. Crawford

Igneous intrusion and metamorphic processes in pre-Cambrian time have been important factors in modifying the structure of the pre-Cambrian terrane whose rocks have been described in previous chapters. Vertical movements have many times effected the submergence and emergence of the region of which the Red Cliff district is a part, and have left the district high above sea level; but in this chapter will be considered chiefly the results of relatively minor movements that have occurred in post-Paleozoic time.

West of the district are Holy Cross and Notch mountains at the north end of the Sawatch Range. The rising of these mountains has tilted the Paleozoic sediments which extend from the mountain slopes to the Mosquito fault (Fig. 12) where the beds abut against the pre-Cambrian rocks. In the mapped area the sediments have a northeast dip of 5° to 15°. In the mines of Battle Mountain the dip is nearly 10°; for the entire district the average dip is about 7° or 8°.

FAULTING

Faults of great displacement seem to be absent from the Red Cliff district. Though many of small displacement exist, they are hard to find in the pre-Cambrian rocks and in the limestone, and can be traced only short distances. Most of those shown in pre-Cambrian rocks on the accompanying claim map (Pl. II) were disclosed in underground workings. These range from small displacements, where evidence of movement is scarcely discernible, to larger faults where the brecciated zone is several feet wide. In several places between Red Cliff and Gilman there has been movement along fissures in the igneous rocks. In these larger faults, where displacement can not be measured because of the character of the wall rock or the limited evidence available, and where the amount of gouge and brecciation is considerable, it seems not unreasonable to suppose that an oscillatory movement has taken place,—that is, the same wall of the fault has moved in opposite directions at different times. In some instances the fault is not a single dislocation, but appears as a number of closely spaced breaks; or a single large fault may fork, and the more mineralized branch be followed by the drift or tunnel. In several places underground evidence of two periods of faulting may be seen where the northeast-striking faults are offset by cross faults of minor importance.

From Gilman to Red Cliff and thence southeastward a mile or more sharp breaks in the outcropping Sawatch quartzite are numerous. The most conspicuous cracks, or fissures, strike northeastward in the direction of dip of the quartzite, and they stand nearly vertical. The quartzite bedding generally does not show any offsetting at these breaks, and most of them might easily be mistaken for dip joints. Rarely a bed may be offset a few inches or, at most, a few feet. Numerous mine workings show many of these breaks to be planes of faulting, particularly those that continue up into the limestone. Broken and shattered rock, slickensiding, and gouge are common. In places fault rock forms a zone up to 10 feet wide. Underground, as at the surface, where there is any stratigraphic displacement it is very slight.

The conditions observed might have been brought about by a more or less vertical differential movement along a plane, the fault blocks settling down into the same relative position that they occupied before the movement; or there may have been differential movement along these planes between blocks that slipped in a direction parallel to the bedding planes. That a differential movement almost at right angles to the bedding did take place is shown by the nearly vertical grooving on fault walls. This grooving was observed in the limestone of the Eagle mines and the Foster Combination mine, in the quartzite of the Ground Hog mine, and in the quartzite of a cliff south of Eagle River near Gilman. Careful search by the Survey party has failed to disclose evidence of vertical faulting in the outcrop of beds overlying the Leadville limestone. Horizontal or nearly horizontal grooving has been seen in the quartzite of the Ground Hog and Bleak House mines. It is evident that fault blocks were subjected to the two kinds of movement mentioned at the beginning of this paragraph.

The name fault implies displacement, and no permanent displacement on most of the planes can be proved. Yet there is abundant evidence that at some time there was pronounced differential movement and hence displacement over extensive surfaces. Those planes or zones along which movement was effective in slickensiding and crushing the rock will be hereinafter designated faults, even though they may show no offsetting of beds. Locally certain faults appear to pass into a fault zone several feet wide where the movement was distributed over many closely spaced fractures. This is more easily seen in the outcropping quartzite than in the mines. In places the movement on each plane was so slight that the zone would be unimportant did it not pass gradually into a single fault plane along which movement was concentrated. The dip faults—those that strike approximately in the direction of dip of the beds—are oftener seen in the quartzite than in the overlying limestone, owing perhaps to the greater brittleness of the quartzite. In the deeper mine workings in limestone a fault apparently may be locally scattered through a minutely fractured zone that shows practically no indications of movement, or it may fade out entirely.

Besides the dip faults there are on Battle Mountain faults of high dip that strike northwestward and others that strike nearly due west. Most of these are unimportant, but in the mines of the Empire Zinc Company are two along which there has been much movement. While most or all of the dip faults were made before the ore was deposited it is possible that one northwest striking fault is younger than ore.

Bedding faults where one bed has slipped over another along a plane parallel to the bedding are common in the mines, but are rarely detected at the surface. These faults commonly show a layer of gouge from a fraction of an inch to 8 inches thick, mostly less than 2 inches. Locally a fault may cut across the limestone and continue along a bedding plane only a few inches above or below the one it left. Generally there has been very little crushing of the rock along the bedding faults, and the displacement may be little. In addition to the bedding faults and those of high dip are others, seen underground, that have intermediate dips and no uniformity of strike.

Between Red Cliff and the south border of the mapped area are many faults of small displacement and northeastward strike. These can be seen in the Cambrian quartzite and jasperoid (p. 56) that has replaced the faulted Leadville limestone. None can be traced far on the surface. It is probable that several faults are covered with landslides through part of their course, owing to deep erosion along the less resistant material of the fault zones. Many faults that preceded the extensive siliceous replacement of the limestone (p. 58) have been preserved in the replacing jasperoid.

About three-fourths of a mile west of the mouth of Silver Creek the quartzite on the northwest side of a fault has been raised an undetermined distance and tilted; it now has a northwest dip of about 75°. Owing to mantle rock and jasperoid which cover the fault the exact strike cannot be determined, but it is probably about N. 40° E. Considerable limonite was found southeast of the fault, or in the fault zone, in prospects now caved. Nearly in line with this fault, on the west bank of Eagle River, there has been faulting in the gneiss; this is seen in a very small exposure. East of the river, still in line with the fault, is a landslide 400 or 500 feet wide (Pl. I.) Evidence points to channel cutting along a fault line, and subsequent slumping. The known fault strikes southwest toward Homestake Gulch, and many have determined the position of the gulch. Between the gulches of Homestake Creek and Eagle River the highly resistant jasperoid prevented rapid erosion along the fault.

Between 1 and 1.5 miles southeast of the fault just described are several faults in the quartzite. At a few a displacement of several feet is indicated. One, seen at the railroad and shown on Plate I, dips 50° southeastward. On the southeast side the quartzite has been dragged up, and now dips southeast. On the hill toward the southwest and almost in line with the fault at the railroad the jasperoid has preserved a fault that dips 76° N. 23° W. The offsetting of the quartzite-jasperoid contact indicates a downthrow on the north side of the fault. Across the river, toward the northeast, is one of the largest landslides of the district in an area probably made weak by faulting.

A small fault shows in the steep slope a short distance northeast of Pando. Several have been preserved by the jasperoid on both sides of Elk Creek.

About a mile and an eighth south of Deen Station is a fault in the Cambrian quartzite and pre-Cambrian rocks. The downthrow is on the north side and the displacement is about 80 or 90 feet. Owing to talus along the line of faulting the direction and amount of fault dip has not been determined. The displacement at this fault is the greatest noted in the district.

From a structural standpoint faults in the Red Cliff district are unimportant. Their importance in connection with ore deposition is brought out in Chapter VI.

CHAPTER VI
MINERAL DEPOSITS[13]
By R. D. CRAWFORD

In this chapter both ore minerals and minerals of no commercial value will be treated. The valueless minerals form masses that were deposited by large-scale geochemical processes that were evidently more or less directly involved in ore deposition.

GENERAL CHARACTER OF THE ORES

Gold, silver, copper, lead, zinc, and manganese, in commercial quantities, have been produced in the Red Cliff district. The early production was chiefly from the oxide zone; it included gold, silver, and lead. In recent years most of the ore has come from the primary-sulphide zone. The metals produced from sulphide deposits are zinc and silver in large quantities, besides considerable gold and copper; comparatively little lead sulphide has been produced in recent years. Secondary sulphides are unimportant in this district. Native gold—part of the ore very rich—has been found in pyrite in the quartzite and in iron oxide derived from pyrite. Many carloads of manganese oxides have been shipped from Bell's Camp. From Table I, page 12, it is seen that the average annual value in recent years of ore shipped from the district ranges from about $17.50 to $30.00 a ton. The figures below are taken from the record of smelter settlement sheets for many thousand tons of ore shipped, during the years 1908 to 1918 inclusive, from several mines operated by Dismant Brothers and others. Where not otherwise stated the items are for carload lots, or at least lots of several tons each.

Range in ore constituents

Gold: nothing to 22.8 ounces per ton.
Silver: 2.4 to 306.2 ounces per ton; one lot of 183 pounds from the Ground Hog mine yielded 303.4 ounces gold and 7,671.2 ounces silver per ton.
Lead: nothing to 56.6 per cent, mostly low.
Iron: nothing to 64 per cent.
CaO: nothing to 28 per cent.
SiO_2: 3.2 to 88 per cent.
H_2O (moisture): 1 to 46 per cent, mostly below 12 per cent.

The ores with highest lime content came from mines in the Leadville formation, and those with highest silica content came from mines in the quartzite. In both cases it is evident that the ore minerals had replaced only part of the rock, perhaps near the border of the ore body.

SULPHIDE ENRICHMENT

Secondary-sulphide minerals in the Red Cliff district appear to be nearly negligible in quantity. Mr. I. A. Ettlinger, who has studied the geology and ores of the Eagle mines in detail, makes the following statements: "The only sulphide enrichment observed was the black coating on the pyritic ore. This coating is found mainly on the ore running high in copper, being a

---

[13]Most of the laboratory tests of ore and gangue minerals collected by the Survey party were made by Russell Gibson and E. A. Hall. Mr. Gibson also wrote the paragraph on fissure veins.

replacement of the chalcopyrite by covellite or chalcocite. Pyritic ore carrying this coating runs high in silver." Mr. Hall has found a little sulphide enrichment in the Ground Hog mine described in Chapter VII. That only slight secondary-sulphide enrichment has occurred may perhaps be explained by the rapid flow of ground waters from the veins into the deep canyon of Eagle River, and the consequent loss of dissolved metals.

MINERALS OF THE ORE DEPOSITS

In this section are noted only the characters shown by the ore and gangue minerals of this district. Full descriptions of the same minerals are to be found in many mineralogy text-books. The metallic minerals are listed alphabetically under the name of the most important contained metal. For convenience, the formula of each mineral is given, together with the percentage of the principal-metal content of the pure mineral.

Mr. A. W. Pinger has made for the Empire Zinc Company a thorough examination of the ores of the Eagle mines. Polished ore specimens were examined with the aid of a metallurgical microscope, and the microscopic work was supplemented by numerous assays and analyses by J. D. Hawthorne, chemist for the same company. The officers of the Empire Zinc Company generously permitted the writer to read the report on the ores and quote therefrom. The statements credited to Mr. Pinger in the following paragraphs are taken from the report mentioned.

ALUMINUM

Alunite, h y d r o u s potassium-aluminum sulphate, $K_2O.3Al_2O_3.4SO_3$. $6H_2O$. Friable white alunite was found associated with zinc carbonate in a zinc-ore stope in the Eagle No. 2 ore shoot. The one sample taken carries some zinc, calcium, and magnesium.

ARSENIC

Arsenopyrite, iron arsenosulphide, FeAsS—iron 34.3 per cent, arsenic 46.0 per cent, sulphur 19.7 per cent. Mr. Pinger has found minute quantities of this mineral associated with pyrite from the Eagle mines.

COPPER

Bornite, copper-iron sulphide, Cu_5FeS_4—copper 63.3 per cent. "Bornite was observed in four specimens. In these it occurs in very minute veinlets in chalcopyrite, visible only with the aid of the high-power objective. In one specimen there is a small area apparently intergrown with chalcopyrite. In this instance it may be contemporaneous with the chalcopyrite; but in the other observed occurrences it is probably a secondary product" (Pinger).

Chalcanthite, hydrous copper sulphate, $CuSO_4.5H_2O$—copper 25.45 per cent. Sky-blue chalcanthite fills a few narrow fissures in the quartzite in the Ground Hog mine. Some specimens collected in this mine show pyrite and chalcanthite in close association. A blue stain is common on the walls of other mine workings in the quartzite. This is probably chalcanthite formed through oxidation of sulphide ore since the mines were opened.

Chalcocite, cuprous sulphide, Cu_2S—copper 79.8 per cent. Black, granular chalcocite was seen in small quantity by Mr. Hall in the Ground Hog mine.

Chalcopyrite, copper-iron sulphide, $CuFeS_2$—copper 34.5 per cent. Specimens of this mineral were collected by the Survey party from the Ground Hog and Pursey Chester mines. In the latter mine practically pure chalcopyrite is associated with pyrite, zinc blende, and quartz. The same mineral is said to have been produced in considerable quantity by the Iron Mask mine. It probably furnished much or most of the copper that has been produced by the district in recent years. Of ore from the Eagle mines Mr. Pinger says: "Chalcopyrite is generally present in the ores in very small quantities. In many of the specimens studied it was entirely lacking. A few polished specimens contain nearly three-quarters chalcopyrite. It is rarely present in the ores in quantities large enough to be readily seen in the hand specimen. Under the microscope chalcopyrite occurs surrounding, filling in between, and sometimes corroding grains of pyrite. The chalcopyrite is often intergrown with sphalerite, and commonly appears to have been corroded and partially or completely replaced by sphalerite; small residual specks often occur in the sphalerite and have an arrangement roughly parallel to the irregular contact between the two minerals."

Covellite, copper sulphide, CuS—copper 66.5 per cent. "Covellite is present in very small amounts, but rather more widespread than bornite. When present it occurs as a thin dark-colored film or coating on the surface of cracks and open spaces in pyritic ore, and is usually restricted to ore containing chalcopyrite. Under the microscope covellite is observed cutting and partly replacing chalcopyrite along tiny veinlets. From these relations covellite is probably to be considered as of secondary origin,—that is, it is copper sulphide that has been leached out of overlying ores by downward circulating ground or surface waters, and reprecipitated below the zone of oxidation by partial replacement of chalcopyrite" (Pinger).

Tetrahedrite. copper sulphantimonite, $Cu_8Sb_2S_7$—copper 52.1 per cent. Of tetrahedrite and freibergite Mr. Pinger writes: "These minerals occur in very small amounts in those ores which have comparatively high silver values. The freibergite is distinguished from tetrahedrite by the presence of silver; under the microscope freibergite reacts slightly to dilute nitric acid and tetrahedrite is negative to this micro-chemical test. The two minerals are very similar in appearance and in occurrence and associations. In many instances it was impossible to determine which of the two was under observation. They are usually associated with chalcopyrite and pyrite, and are occasionally corroded and partly replaced by sphalerite, and to a less extent by chalcopyrite."

GOLD

Gold commonly occurs in sulphide and oxidized ores found in the quartzite and pre-Cambrian rocks. In the high-grade ores of the Ground Hog, Champion, and possibly other mines in the quartzite native-gold nuggets were frequently found. It is said that nuggets up to 2 ounces in weight were not rare. The nuggets are reported to have been angular and horn shaped in the pyrite, and more or less smoothed and rounded in the oxidized ore.

IRON

Coquimbite, iron sulphate and water, $Fe_2(SO_4)_3.9H_2O$—iron 19.87 per cent. This mineral was found on the walls of a crosscut in the Eagle No. 2 ore body where it evidently had formed by oxidation of the iron sulphide after the ore was blocked out. The mineral occurs in aggregates of poorly formed crystals. A green color predominates; patches are bluish green. Part of the material reacts only for ferric sulphate and water, though much of it carries a small amount of copper. Whether copper replaces part of the iron or occurs in chalcanthite mixed with the coquimbite has not been determined.

Limonite, hydrous iron oxide, $2Fe_2O_3.3H_2O$—iron 59.8 per cent. Brown and yellow limonite is one of the commonest minerals in the oxidized zone, and is frequently found far from any known ore body. Much of the limonite is admixed with kaolin and other materials.

Pyrite, iron disulphide, FeS_2—iron 46.6 per cent, sulphur 53.4 per cent. Pyrite occurs in nearly or quite all the sulphide ore bodies in the district. Though in places it has no value as ore, it generally carries gold, silver, or copper. Pyrite is the best silver ore of the deep mines in the limestone. Much of the gold and silver of the mines in the quartzite occurs in pyrite. Well formed crystals—cubes or pyritohedrons, or combinations of the two—are common. They range in size from one-sixteenth inch to an inch or more in diameter; the largest crystals seen by the writer were found in the quartzite. Most of the pyrite of large bodies is massive, and shows crystals only in drusy cavities. Parts of a body may be made up of porous pyrite where the mineral occupies about three-fourths to seven-eighths of the volume. Most of the open spaces are narrow and less than half an inch long. The walls of the spaces are crystallized pyrite.

Pyrrhotite, iron sulphide, about $Fe_{11}S_{12}$. Of this mineral Mr. Pinger writes: "Pyrrhotite is present sparingly and has been observed only in extremely minute residual grains in sphalerite and galena. In fact it occurs in such small quantities that the micro-chemical tests were somewhat indefinite."

Siderite, iron carbonate, $FeCO_3$. Light gray to brownish siderite is one of the most abundant gangue minerals of the district. Cavities in siderite masses show well shaped simple rhombohedrons, and the massive mineral is coarse in texture. Manganese and magnesium replace a large part of the iron as shown by the nearly complete analyses below. The sulphur comes from associated pyrite. The two specimens analyzed came from different parts of the mines.

Analyses of siderite from Eagle Mines

(P. M. DEAN, ANALYST)

	1	2
FeO	35.73	25.07
MnO	11.30	12.29
MgO	10.56	16.13
CaO	.87	1.75
CO_2	40.10	42.50
S	.35	.18
Insoluble	.26	.17

Turgite, ferric oxide with water. Red earthy turgite is found in several mines and prospects in the oxide zone.

LEAD

Anglesite, lead sulphate, PbSO₄—lead 68.3 per cent. According to S. F. Emmons[14] much anglesite in minute crystals, or "sand," was found in the oxide zone of the mines in the Red Cliff district.

Cerussite, lead carbonate, PbCO₃—lead 77.5 per cent. This mineral was common in ores shipped from the district in early days. It is now found in small quantities on the borders of old stopes. The Survey party collected only a few samples of lead carbonate, most of which is soft and carries an admixture of iron oxide. Samples assayed show 9 to 12 ounces silver and .05 to .07 ounces gold per ton. With the soft earthy carbonate are small groups of cerussite crystals having adamantine luster.

Galena, lead sulphide, PbS—lead 86.6 per cent. In mines now operating galena is found only in small quantity. That seen by the field party occurs in grains and crystals up to half an inch in diameter. The crystals are combinations of cubes and octahedrons. In the deepest mines in the limestone the galena carries little or no silver. A sample of galena from the quartzite in the Ground Hog mine, assayed by Mr. Hall, yielded .72 ounces gold, and 34.30 ounces silver per ton. A second sample from the same mine assayed .15 ounces gold and 17.92 ounces silver per ton.

MANGANESE

Oxides of manganese are common in the oxide zone, and a large tonnage of manganese ore is reported to have been shipped from Bell's Camp several years ago. Specimens of manganese ore collected are mostly mixtures of pyrolusite and psilomelane. Some of the material carries iron, and much of the black material is manganese-bearing limonite. In the mines and on the dumps were found pseudomorphs of oxides of iron and manganese after siderite. It is probable that most of the manganese ore has been formed by the alteration of siderite.

SILVER

A large part of the silver produced by the oxidized ore bodies is said to have been combined with chloride in the mineral cerargyrite, or horn silver. Samples collected from the Rocky Point and Polar mines contain a little cerargyrite that can be detected only by chemical tests. Native silver and argentite have been reported from the district. Mr. Pinger has found a small amount of freibergite (silver-bearing tetrahedrite) associated with pyrite and chalcopyrite. Richard Pearce[15] states that hessite (silver telluride) was found in large quantities in one or two of the mines at Red Cliff.

ZINC

Goslarite, hydrous zinc sulphate, ZnSO₄.7H₂O—zinc 22.73 per cent. White zinc sulphate is found in old stopes of the Eagle mines near the top of the sulphide zone. The mineral was formed after air reached the sulphide ore through mine openings; it is well preserved in dry parts of the mine. This material lines one of the Iron Mask haulage ways where it protrudes through the openings between lagging, and hangs like a mass of soft white hair in fibers a foot long. Qualitative tests show that a little iron is present with the zinc.

---

[14]Colo. Scientific Society Proceedings, vol. 2, 1886, p. 101.
[15]A. I. M. E. Trans., vol. 18, 1890, p. 541.

Smithsonite, zinc carbonate, $ZnCO_3$,—zinc 52.2 per cent. Hard, massive, gray zinc carbonate is found on the borders of a few old stopes. Some of it carries a little calcium and magnesium. One sample of red carbonate ore collected contains considerable copper and iron as well as zinc. Qualitative tests of the gray carbonate show much water, thus indicating that hydrozincite is present with the smithsonite.

Zinc blende, or sphalcrite, zinc sulphide, ZnS—zinc 67 per cent. This mineral has made practically all the zinc ore that has been produced in large quantity by the Eagle mines. In these mines the zinc blende is dark brown to black owing to a considerable percentage of iron and is the variety known as marmatite. It is partly massive and solid, and partly friable. Much of the best ore is friable and sandlike. The best bodies of zinc ore are nearly free from other minerals, but some of the ore shows an admixture of galena, pyrite, or siderite. Drusy cavities contain numerous small blende crystals, most of them not more than one-sixteenth inch in diameter. In the Ground Hog mine blende, too scarce to make zinc ore, occurs in crystals up to half an inch in diameter. Most of the blende of this mine is also dark brown. A lighter variety, "rosin zinc," is found in small amount in a few mines of the district. Zinc blende is found in small quantity in the deepest mines in the pre-Cambrian rocks.

EXCLUSIVELY GANGUE MINERALS

Barite, barium sulphate, $BaSO_4$. Barite as a gangue mineral in this district is not common, but in several fracture zones patches of small barite crystals form a coating on the rock at or near the outcrop. North of Elk Creek is a narrow vein of coarse, cleavable, massive barite.

Calcite, calcium carbonate, $CaCO_3$. Crystallized calcite is found in cavities in the oxidized ore where it has evidently been deposited by cold surface waters.

Dolomite, calcium-magnesium carbonate, $CaMg(CO_3)_2$. In addition to the fine-granular and dense dolomite and dolomitic limestone of the Leadville formation that incloses many of the ore bodies, coarse-granular dolomite is common near the ore. This coarse dolomite has evidently replaced fault rock, including gouge. Much of it is medium gray in color, but in several places it is mottled and shows angular patches of white and very dark gray. The dark gray dolomite carries minute crystals and grains of pyrite. After the carbonate and pyrite are dissolved in acids, the small amount of residue—still dark gray—seems to be composed chiefly of quartz and carbonaceous material. Nearly complete analyses of the dolomite are given below.

Analyses of coarse dolomite from Eagle Mines
(P. M. DEAN, ANALYST)

	1	2	3
CaO	30.28	30.10	29.41
MgO	19.72	20.18	20.48
FeO	.78	1.85	1.22
MnO	----	.42	.44
CO$_2$	45.09	46.20	46.17
S	.20	----	----
Insoluble	2.33	.72	1.30

1. Gray dolomite, Wilkesbarre shaft, intermediate level.
2. White dolomite from mottled white and dark gray mass near ore body, Eagle No. 2.
3. Dark gray dolomite in contact with white dolomite of sample No. 2.

Gypsum, hydrous calcium sulphate. Gypsum in small quantity can be detected by chemical tests in the earthy material of the oxide zone in many mines and prospects. Pieces of transparent crystallized gypsum (selenite) are found in the Potvin mine, where this mineral is associated with pyrite, limonite, and oxide of manganese. Small crystals of selenite form a shiny coating on masses of iron and manganese oxides in the stopes of the old Black Iron mine, now Eagle No. 2.

Kaolin, hydrous aluminum silicate, $Al_2O_3.2SiO_2.2H_2O$. White soft kaolin, commonly associated with quartz, is found in the veins and on the vein walls of most of the mine workings in the pre-Cambrian rocks, where it has been derived from alteration of the feldspars of the wall rock. The decomposition of the porphyry that lies just above the largest ore shoots in the limestone has resulted in the formation of much kaolin.

Quartz, silica, SiO_2. Ordinary vein quartz, both massive and in small crystals, is found in many mines and prospects; but quartz in these forms is not an important gangue mineral in the ore shoots within the Leadville limestone. Quartz of the quartzite walls may be considered a gangue mineral of some low-grade ores replacing quartzite. To what extent quartz in the jasperoid form was present near the oxidized ores of the early workings is unknown. Further reference is made to this in the section on "replacement ore deposits."

CLASSES OF MINERAL DEPOSITS

All the known ore of the Red Cliff district is younger than the rocks that inclose it. The two following classes are present: (1) fissure-vein ore in the pre-Cambrian rocks and Sawatch quartzite; (2) replacement deposits in the Leadville formation and in the Sawatch quartzite.

Fissure Veins

Most of the fissure veins in the pre-Cambrian rocks strike northeast, and are vertical or have high dips to the southeast. Movement along the fissures prior to ore deposition has resulted in gouge and breccia up to 30 inches in width. Where slipping has occurred along closely spaced breaks the fault zone may be as much as 6 feet wide. The vein minerals are quartz, pyrite, sphalerite, galena, and probably a little chalcopyrite. The pyrite, much of which is copper-bearing, commonly carries both gold and silver. The veins of ore are thin, many being each less than an inch thick; a foot of solid sulphides is rare. Commonly one thick vein splits into two or several which, farther on, may coalesce. The associated decomposed wall rock contains so little disseminated ore that the widening of veins by replacement of the country rock is not considered important. Although there have been two periods of fissuring, positive evidence of post-mineral faulting has not been seen by the writer. In general, pyrite, the most abundant sulphide, was one of the first minerals deposited; sphalerite and galena were deposited later.

Fissure veins in the pre-Cambrian rocks pass into the overlying quartzite where they inclose workable ore bodies. Other good veins worked in the quartzite carry little or no ore in the rocks below. Many veins in the quartzite are narrow, and in places they show distinct crustification. More commonly there has been replacement of the wall rock, and the veins may be considered transitional between the replacement bodies described below and the fissure veins.

REPLACEMENT ORE DEPOSITS

In this district nearly all the ore found in limestone and the greater part of that found in quartzite is of the replacement type. By this it is meant that the ore was brought in by the same solvent that carried the country rock away from the place taken by the ore. Both deposition of ore and removal of rock are believed to have been effected at the same time by exchange of particle of ore for particle of rock. It is possible that narrow fissures existed in places in the limestone along the faults, and that such fissures were afterward filled with ore; but since the replacement extended far into the wall rock any slight fissure filling is negligible in comparison with the whole. In the Sawatch quartzite, owing to less easy solubility of wall rock, fissure veins are relatively much more important than in the limestone.

The large bodies of sulphide ore blocked out in the Eagle mines in 1921 afforded excellent opportunity to observe the results of replacement. Evidences of replacement noted in the Eagle and other mines of the district are the following: (1) irregular shape of the ore bodies; (2) frequent occurrence of perfect pyrite crystals in limestone, quartzite, and gouge near workable ore and, less commonly, in the pre-Cambrian rocks near the veins; (3) presence of unsupported masses of rock in the ore; (4) preservation by the ore of original bedding planes, rock fractures, outlines of disturbed limestone blocks, and structure of fault rocks including faults themselves.

SHAPE OF ORE BODIES

Although the more extensive ore shoots generally follow the northeastward striking faults (dip faults) and pitch approximately with the dip of the beds, their outlines are irregular in that bulges and tongues of any shape may extend many feet on either or both sides of the fault. The smaller bodies may assume any form, and some have their longer dimensions nearly parallel to the bedding planes of the rocks.

UNSUPPORTED ROCK MASSES IN ORE

The sulphide-ore bodies of the Eagle mines inclose occasional small patches of gouge that have escaped complete replacement. The patches are likely to carry some pyrite near their borders. Near the top of ore shoots are seen what might be either shale or gouge along a former bedding fault. Such a streak 2 inches thick was traced 10 feet nearly horizontally on the 7th level of the mine. That these patches have not worked into the ore as a result of faulting after ore deposition is proved by the solid and unfaulted character of the sulphide that surrounds them. Mr. Hall has noted blocks of quartzite surrounded by ore in the Ground Hog mine.

CRYSTALS IN COUNTRY ROCK

Pyrite crystals having part or all of their faces developed occur in limestone and quartzite near the ore bodies and in the wall rock of the fissure veins in the pre-Cambrian rocks. The fault rock, including gouge, locally carries perfect crystals of pyrite. The crystals are commonly less than one-sixteenth of an inch in diameter. Cubes and pyritohedrons are the commonest forms. These disseminated crystals in places are rare; in other places they make up 10 to 20 per cent of the rock volume. Evidently they have been precipitated from solutions that penetrated the rock beyond the limits of workable ore.

PRESERVATION OF ROCK STRUCTURES BY ORE

Since most of the limestone is thick bedded or massive, having few or indistinct bedding planes, it is not to be expected that indications of bedding planes will be common in the ore that replaces the limestone. A very fine-grained bed of the quartzite member of the Leadville formation is laminated and locally shows minute fractures, faults, and folds that were evidently impressed on it before cementation was completed. All these structural features are preserved in the pyrite of the Eagle No. 2 ore shoot in about the position where one would expect to find the quartzite had it not been replaced. In the No. 1 ore shoot of the Eagle mines, in a stope below the 14th level, lines of original rock bedding were seen on the ore face.

More conspicuous examples of preservation of former structures are seen where the ore has replaced fractured and faulted rock. The phenomenon is best shown by pyrite or by mixed pyrite and zinc blende. In pure zinc blende the details of former structure are not so evident. A good example of replaced shattered limestone was seen near the east side of the No. 1 ore shoot of the Eagle mines on the 14th level. Here blende-pyrite fragments and blocks from a fraction of an inch to several inches in diameter made the vertical wall of a drift. This wall, or ore face, was self-supporting because the component blocks were interlocking and tight fitting; they had not been subjected to movement or crushing since the ore was formed. Yet, so loosely were the fractures of the former limestone cemented, the sulphide blocks could be pried out with a pick. When this and similar faces of ore are covered with dust they can not be distinguished from limestone by surface appearance alone. On the 6th level of Eagle No. 2 pyrite preserved through a zone 7 feet wide all the structural details, except slickensiding, of a former fault. In this place the blocks were firmly bonded, probably owing to the presence of gouge before replacement. This fossil fault is more fully described on page 61.

SIZE OF REPLACEMENT ORE BODIES

The greatest length of most of the replacement ore bodies lies in the dip of the sedimentary beds and along fault lines, that is, toward the northeast with a pitch of about 10°. The largest known body has been developed continuously through a distance of 3,060 feet, all in the dolomitic limestone. A second body has a known length of nearly 3,000 feet. Others are known to be more than 1,000 feet long.

Part of the ore shoots are elliptical in cross-section, having the longer diameter of the ellipse nearly horizontal. One of the largest shoots is 50 to 100 feet thick and 75 to 150 feet wide (Pl. III, in pocket). A variation from the above type is the blanket vein, which is parallel to the bedding of the inclosing rocks and commonly thinner than the shoots with elliptical cross-section. The largest known blanket-vein deposit in the Leadville formation is 6 to 50 feet thick and 75 to 150 feet wide. In the underlying quartzite the blanket veins are said to have been 2.5 to 16 feet thick; in width these veins were commonly less than 15 feet though locally more than 100 feet. Replacement bodies of very irregular shape have been found, but almost invariably their longest dimensions were nearly parallel to the bedding planes of the inclosing rocks. Other replacement bodies have roughly plane-parallel vertical walls, and have been formed by replacement of the walls of the fissures through which the ore-bearing solutions came.

SIDERITIC AND DOLOMITIC REPLACEMENT

Siderite, carrying magnesium and manganese, is very common immediately under and at the sides of the ore bodies; it is less plentiful above the ore. The siderite layer ranges in thickness from less than an inch to 25 feet. Analyses of siderite from the Eagle mines are given on page 49.

Below the siderite, and in places many feet from ore, are large masses of coarsely crystalline dolomite. Smaller masses are found overlying ore. This dolomite is commonly bluish gray, but in places many small white patches are found with dark gray patches. Many of the white patches are angular, and locally a mass suggests the replacement of former fault breccia by dolomite, white patches in the place of fragments and gray in place of matrix. The chemical composition of the white and gray dolomite is almost identical as shown by analyses Nos. 2 and 3 on page 51. Chemical tests show a small quantity of carbon after all the carbonate has been removed. This carbon is probably unreplaced carbon of the original gouge or shale, and to it the dolomite may owe its dark color.

That the dolomite, like sulphides and siderite, replaces former fault rock is indicated by the following: (1) the coarsely crystalline dolomite is commonly found under ore near the large ore shoots which demonstrably follow faults. (2) On the intermediate level of Eagle No. 1, near the shaft, specimens of dolomite were collected that had preserved fractures of the former rock and grooved slickensided fault surfaces lacking only the original polish; yet the dolomite is solid and shows no indication of breaking since it crystallized. One specimen from the same place carries a thin streak of gouge partly replaced by dolomite. (3) On the 6th level, below Eagle No. 2 ore shoot, the back and walls of a newly driven drift were composed of mottled white and gray coarse dolomite; many of the white patches were sharply angular. At the breast was a mass of black shale with bedding planes disturbed and twisted by faulting, a small amount of the shale having been replaced by pyrite. · Some of the fault rock was composed of angular black shale fragments cemented by coarse dolomite (mostly gray) that had evidently replaced the original matrix; here the process had stopped before the difficultly replaceable carbonaceous shale was appreciably affected.

Coarse gray dolomite in many places extends a considerable distance into dolomitic limestone very similar in chemical composition to the coarse dolomite. Since the limestone was originally massive no preserved structural features have been detected outside of a former fault zone. The coarse dolomite is found in large masses in a few places, at the outcrop and in prospects, far from any known ore deposit. It is probable that this dolomite replaced rock in and near faults at the same time and in the same manner that the dolomite was deposited near the sulphides. From the facts noted above it follows that this dolomite is in itself evidence of faulting in its vicinity, but an actual fault in the unreplaced underlying quartzite proves this relationship where the dolomite outcrops about three-fourths of a mile north of Gilman.

JASPEROID REPLACEMENT

Though the largest jasperoid replacements in this district occur far from any known ore body they are described in this place because of their evident relation to faults and dolomitic replacements and because of their probable relation to the sulphide deposits. The name jasperoid was given to this kind of quartz by Spurr[16] who described similar replacements in the Aspen district.

DESCRIPTION OF THE JASPEROID

The jasperoid is composed chiefly of microgranular to cryptocrystalline quartz, and varies from dark gray to neutral gray to yellowish- and brownish-gray. Most of the rock feels gritty, and breaks with uneven to subconchoidal fracture. Another variety resembles flint, and has a good conchoidal fracture. Both kinds show small cracks and cavities that were probably formed by shrinkage either in the replacement process or at the time of crystallization. In the dense variety cavities may have a diameter up to one inch. In both kinds the openings form only a small fraction of the rock volume.

The densest jasperoid is in part finely banded or streaky. The streaks are less nearly parallel than are the bands of agate. The microscope shows the darkest streaks to be composed of cryptocrystalline or nearly amorphous quartz with a large proportion of black carbonaceous material. The lighter streaks are composed chiefly of fine-microgranular quartz with less black material. Throughout the thin section in the lighter streaks are a few grains of quartz which, though very small, are many times larger than most of the grains. The rock carries a few minute flakes of muscovite. This variety somewhat resembles phases of the Tintic (Utah) jasperoid described by Lindgren[17] who interprets the silicification as "a replacement of limestone or dolomite by colloidal silica which immediately afterwards became transformed into chalcedony or in part into granular quartz." A similar origin for the densest jasperoid of the Red Cliff district is further suggested by the shape and appearance of cavities that were evidently formed as a result of shrinkage and settling of the gelatinous mass during the process of solidification and crystallizaton of the silica.

Most of the coarser variety is microcrystalline, though locally it resembles very fine-grained quartzite. In places the small cavities are lined

[16]Spurr, J. E., Geology of the Aspen mining district, Colorado: U. S. Geol. Survey Mon. 31, 1898, pp. 216-221.
[17]Lindgren, W., Processes of mineralization and enrichment in the Tintic mining district: Economic Geology, vol. 10, 1915, pp. 225-240. See also description by the same author in U. S. Geol. Survey Prof. Paper 107, 1919, pp. 154-158.

with minute quartz crystals. Some specimens are uniform in texture; others have angular patches of slightly coarser grain than the matrix. In a thin section examined limonite, which colors the rock, is more plentiful in the matrix than in the angular patches. While it is possible that brecciation occurred before the process of solidification was completed it seems more likely that the jasperoid has replaced a formerly brecciated rock.

The quartz grains of the coarser variety are irregular and without crystal outlines. In some specimens the grains are equidimensional; in others they are elongated or elliptical in outline, having one diameter two or three times as long as the other. A few thin sections carry occasional larger grains. Of ten thin sections examined none shows quartz with crystal outline. In this feature the rock is unlike phases of the jasperoid at Aspen described by Spurr[18] and by Lindgren[19].

Most of the jasperoid is remarkably free from residual limestone or dolomite. Careful search was made for phases that might show only partial replacement, but the search was unsuccessful at most of the outcrops examined. However, two thin sections were made of specimens that contain carbonate. One of these carries anhedral quartz grains among the dolomite grains in the less siliceous part of the specimen. In the more siliceous part are a number of fairly large shapeless grains and also roughly rectangular grains of quartz inclosing a few crystals and anhedrons of dolomite, while many grains of dolomite occupy the spaces among the quartz grains. Most of the second thin section is streaky and like the densest jasperoid described above. The siliceous part ends abruptly in a wavy line against the unaltered dolomite.

OCCURRENCE OF JASPEROID

Masses of jasperoid many feet in diameter replace Leadville limestone at the outcrop near Turkey Creek and thence southward at intervals for about 2 miles. A large body is found about three-fourths of a mile north of Gilman above the public road. Jasperoid was not detected by members of the Survey party in the mines at Gilman, but it can be seen on mine and prospect dumps at the Iron Mask, Little Chief, and several openings between the Little Chief mine and the Gilman Club House. Blocks up to 18 inches in diameter were seen in this vicinity, and it is probable that bodies of considerable size were formed near the ores.

On the hill between Eagle River and Homestake Creek, where comparatively little limestone remains, there is much jasperoid. Large-scale siliceous replacement is best illustrated northeast, east, and southeast of Pando where through a distance of 4 miles no outcrop of Leadville limestone has been found. Through about half this distance the bedrock is covered, at intervals, by moraines, talus, and slumped material; but in the several exposures only jasperoid and a little quartzite occur between the underlying quartzite and the overlying Pennsylvanian beds. A complete section of the Leadville formation outcrops half a mile north of Silver Creek, and is 156 feet thick. A mile southward from the mouth of Rule Creek the Leadville formation has an apparent thickness of 208 feet; but this may

---

[18]U. S. Geol. Survey Mon. 31, pp. 218-219.
[19]A. I. M. E. Trans., Vol. 30, 1900, pp. 628, 678.

be 37 feet more than the true thickness, owing to a possible repetition of beds along a strike fault of which there are indications. The formation is thus at least 170 feet thick at this point. It is therefore highly probable that limestone having a thickness of about 150 feet has been replaced by silica through most, if not all, of the four-mile stretch mentioned. Some of the largest exposures are represented on the geologic map (Pl. I).

The jasperoid and underlying quartzite together make a cliff north of Coal Creek (Fig. 10) where the jasperoid measures 56 feet from top to bottom, including 9 feet of quartzite. The lower 22 feet of the section mentioned is softer than most of the jasperoid, but acid tests show only a trace of carbonate. The appearance of this part of the section suggests settling and attendant compression. The upper part (34 feet) preserves the original bedding structure and joint planes of the limestone. The quartzite member of the Leadville formation, of which 9 feet remains, has

Figure 10. Jasperoid outcrop north of Coal Creek

Here all the limestone and dolomite of the Leadville formation, having originally a thickness of about 150 feet, has been replaced by silica. The lower part of the outcrop is Sawatch quartzite. In the cliff at the top the thickness of the jasperoid (including 9 feet of quartzite of the original Leadville formation) is 56 feet. Part of the jasperoid may have been removed by erosion.

also been partly replaced by the jasperoid. This is shown by the irregular or curved trend of contact between flintlike jasperoid and quartzite and more certainly by small masses of quartzite suspended in and surrounded by jasperoid. It should not be inferred that the original 150 feet of limestone has been replaced by only about 50 feet of jasperoid. Though there has evidently been shrinkage it is probable that considerable jasperoid has been removed from the top of the cliff by erosion.

In many jasperoid outcrops the faulted or brecciated structure of the replaced limestone has been faithfully preserved. South of Elk Creek, where the jasperoid of the cliff is about 40 feet thick (Fig. 11), six fossil faults were noted. These are all high-angle faults that strike eastward or northeastward. The jasperoid has for the most part replaced broken limestone, but in places the original bedded structure has been preserved. North of Elk Creek is a similar cliff where the jasperoid is at least 40 feet thick. Here the siliceous replacement has preserved details of a strong fault, including original slickensided surfaces; only the polish is lacking.

In only the largest bodies of jasperoid has the structure of unbroken limestone been preserved. The evidence points to replacement of limestone and fault rock by silica from solutions that circulated through the faults. In part of the district the process extended far enough to replace all the unbroken limestone of interfault areas.

Figure 11. Jasperoid outcrop south of Elk Creek
Above the prospect dumps is a body of jasperoid, about 40 feet thick, that has replaced limestone and dolomite of the Leadville formation.

CRYSTALLIZED QUARTZ REPLACEMENT

More remotely connected with the faulting and subsequent mineralization along the faults—if so connected at all—is the limited replacement of Leadville limestone by quartz crystals. About half a mile north of Silver Creek a few quartz crystals up to an inch in diameter were found embedded in the limestone near the top of the formation. Nearer the creek larger crystals, probably from the same general source, were found on the surface of the ground near a jasperoid outcrop. The largest crystal measures 2 by 3.3 inches.

STRATIGRAPHIC POSITION OF ORE BODIES

Ore is found at practically every horizon in the Leadville limestone and underlying quartzite. It thus has a stratigraphic range in these formations of more than 500 feet; in addition, it extends below the quartzite to a depth of at least 250 feet in the pre-Cambrian rocks. (See Pl. III in pocket and Fig. 14.)

Certain beds of the quartzite, because of higher porosity or higher lime content, were more easily replaceable than others, and in them the ore has a much greater horizontal extent than in the more compact pure quartzite where the fissure veins were less widened by replacement. Mr. B. A. Hart states that the lower blanket-vein zone was 160 feet vertically above the bottom of the quartzite, and that 28 feet of quartzite lay between this and a higher blanket vein.

The largest ore bodies at Gilman and nearly all the ore mined at Red Cliff have been found in the upper part of the Leadville limestone just below the overlying shale. One of the Eagle ore shoots (Pl. III), extends from top to bottom of the limestone, and this is said to be the only large ore body of that mine found below the quartzite member of the Leadville formation. In addition to the white lenticular crystalline patches of dolomite of the "zebra lime" (p 36) the uppermost beds of the Leadville formation are slightly more granular than the lower beds. The lower beds, in general, seem to be slightly more siliceous than the upper beds. Qualitative tests of 25 specimens of the Leadville limestone were made by Mr. Hall who found considerable silica in the lower part of the formation. Mr. I. A. Ettlinger has found in the upper half of the Leadville formation a dense bed, 1 to 5 feet thick, that carries silica up to 7 per cent. Most of the ore in the limestone lies above this bed. However, it is doubtful that the much greater volume of replacement in the upper part can be accounted for by texture and chemical composition. It is probable that this zone was most favored because of the overlying shale which dammed ascending mineral-bearing solutions, thus spreading them laterally and diverting their course directly up the dip and immediately under the shale. The first three analyses of the accompanying table are of dolomite collected at the out-crop of the Leadville formation in Eagle Bird Gulch.

Analyses of dolomites from Leadville formation

(P. M. DEAN, ANALYST)

	1	2	3	4
CaO	28.23	29.71	29.50	30.4
MgO	19.48	21.95	21.37	21.9
FeO	.60	.55	.63	
CO_2	43.21	46.84	46.15	47.7
Insoluble	7.47	.24	.30	

1. Dense dolomite, about 30 feet above bottom of Leadville formation
2. Fine-crystalline dolomite, 35 feet below top of Leadville formation
3. Medium-grained crystalline dolomite about 30 feet below top of Leadville formation
4. Theoretical composition of normal dolomite

RELATION OF ORE BODIES TO FAULTS

The large sulphide-ore bodies of the Eagle mines all show a close relation to the faults, particularly to those faults that strike northeastward. The same is true of the smaller sulphide bodies of mines in the Sawatch quartzite and pre-Cambrian rocks. Likewise certain stopes from which oxidized ores have been mined at Gilman and at Red Cliff are in faulted zones; and there is little doubt that all the ores of the district have been more or less connected with pre-existent fractures, fissures, or faults. Owing to more general fracturing of rock near the present outcrop or to more favorable conditions for mineral precipitation, or to both, the oxidized ore near the surface generally extends farther from the principal faults than do the sulphide-ore bodies at greater depth.

Most of the ore shows clearly that it was formed after the faults that strike northeastward, and no ore bodies examined by the writer show signs of having been cut by later faults. However, Messrs. J. M. and R. V. Dismant, who formerly as lessees mined much ore in the Bleak House and Rocky Point mines, found indications of faulted ore in those mines. In the Bleak House mine a blanket vein of zinc blende replacing quartzite was locally offset about 30 inches by a fault. Since the ore has been removed there is nothing to show whether the partly replaced quartzite was faulted before or after the ore deposition. Mr. J. M. Dismant states that a vertical vein of galena cut across the zinc ore body along the fault plane, that the galena was somewhat crushed, and that the ore was slickensided. Post-mineral faulting well may have taken place in several mines of the district, and it is remarkable that so little has come to light. Proof that the ore lies in and near pre-existent faults is based on observations that follow.

In places the faults can be traced from barren ground directly to solid ore. Frequently gouge can be seen in contact with the ore, but since the ore commonly lies immediately under the shale at the top of the Leadville limestone it is not always easy to distinguish between shale and gouge. The presence of gouge in unbroken ore many feet below the shale bed proves conclusively that the ore is in the position of a pre-existent fault. That the deposition was subsequent to faulting is shown by the relationship between ore and gouge; the best example of this was seen in the No. 1 ore shoot of the Eagle mines in the stope below the 14th level where a small patch of gouge was surrounded by solid and unfaulted ore. A specimen taken shows the gouge to be partly replaced by pyrite.

On the 6th level of Eagle No. 2, about 80 feet north of the nearly east west fault and a few feet from ore, a northeasterly striking fault shows in the drift. Blocks of fault rock are partly replaced by pyrite. A specimen from a block of quartzite coated with a thin layer. of slickensided gouge shows both quartzite and gouge to be partly replaced by pyrite crystals. A specimen from a one-inch streak of gouge in the same fault shows 10 or 15 per cent of the gouge replaced by small pyrite crystals. On the same level in the eastward striking fault, blocks of gouge near the ore are partly replaced by pyrite crystals.

On the same mine level in the "chimney shoot" a little south of its center certain features of the pre-mineral faulted rock were clearly evident. Blocks of fault rock, partly angular and partly having surfaces smoothed and curved by friction and movement, were accurately preserved in outline. The only essential feature missing was the polish of once slickensided surfaces. The fossil fault showed through a zone 7 feet wide many nearly perpendicular partings, evidently in planes of former slipping[20].

Excepting some friable zinc blende the sulphide ore of every face examined is solid and shows no crushing or slickensiding. The condition

---

[20]In the Yak mine at Leadville one of the largest ore shoots has similarly replaced faulted rock directly in line with a well defined fault in the unreplaced limestone. This replacement was seen by the writer in the ore of a pillar.

of most of the siderite near the ore is similar, though in one place in the No. 2 ore body of the Eagle mines the siderite presented slickensided surfaces. Even here the siderite tightly adhered over part of these surfaces, thus indicating that they may have been smoothed before the siderite was deposited.

SOURCE OF THE ORES

From the relationships described on preceding pages it is clear that the horizontal distribution of ore bodies was largely controlled by the faults through which the mineral-bearing solutions circulated. The position of the largest ore bodies under the shale and overlying porphyry was in all probability largely determined by the damming of solutions and consequent deposition under the barrier of shale and porphyry. The carbonaceous content of the shale may have aided precipitation. These relationships indicate that the ore was deposited from ascending solutions. Whether the solutions rose vertically or from the northeast or from the southwest is not clear.

The porphyry and ores are probably genetically connected in having emanated from a common deep-seated magmatic source. From what direction did the porphyry magma flow? This is an interesting and important question, though perhaps not vital when considering the *immediate* source of the ores. Even if the intrusion of porphyry preceded the tilting of the beds to their present attitude the comparative scarcity of dikes toward the southwest makes it improbable that the porphyry magma came from that direction. The largest porphyry bodies exposed in the region are in the Tenmile district toward the southeast. Though the Gilman sheet extends many miles southeast and probably is continuous with the Chicago Mountain laccolith, it may have reached its present position by flowing northwest from the laccolithic mass or by flowing west or southwest from a possible vent much farther north than Chicago Mountain. Careful search has failed to disclose good evidence for a northeast source for the porphyry, and the rock probably reached its present position by flow toward the west or northwest. It is evident that mineral-bearing solutions could have flowed in the same direction along contacts and through porous quartzite beds, and thence upward through fault fissures. Such directions of movement for the bulk of the solutions necessitate also their downward movement into fissures in the pre-Cambrian rocks where ore and gangue minerals were deposited. The latter consideration makes it appear unlikely that the solutions came from the east or southeast.

The primary sulphides of the higher levels in the limestone have a much higher proportion of zinc and a much lower gold content than have the primary sulphides in the underlying quartzite and pre-Cambrian rocks. Though galena is relatively scarce in the present workings in limestone the large amount of cerussite and anglesite formerly produced from the oxide ore bodies points to much galena in the primary ores of the higher levels. The zone theory of ore deposition formulated by Spurr[21], and found to hold generally for many districts where the ore has been deposited from ascend-

[21]Spurr, J. E., A theory of ore-deposition: Economic Geology, vol. 2, 1907, pp. 781-795.

ing solutions—probably magmatic—supposes that among others gold and pyrite, copper-bearing pyrite, and galena-blende ores are successively deposited with increasing distance from the magmatic source. This order roughly holds with successively higher zones in the Red Cliff district, though the data are less complete than one might wish.

No indications of a southwest source for the solutions have been noted. The great length of the ore shoots, pitching with the dip of the beds and replacing fault rock, points to (a) upward flow from the northeast or (b) vertically upward flow through a wide northeast-southwest range or (c) upward flow through a more restricted range with flaring path above the vent. In each case it is assumed that the solutions came from considerable depth through the fault fissures. Geologists of the United States Geological Survey have shown that the greater part of the porphyry of the Aspen, Leadville, Alma, and Tenmile districts was intruded prior to the most intense faulting. There is not complete agreement, however, among geologists and engineers concerning the amount of faulting that preceded and the amount that followed ore deposition in the Leadville and Alma districts. ·The State Survey party has found no evidence that determines whether the porphyry intrusion preceded or followed faulting in the Red Cliff district.

Known porphyry or its continuation at depth has been considered the probable source of the metals in central Colorado mining districts by several workers in those districts, but others have supposed that the metals came from magma at greater depth. In a recent articles the writer[22] gives evidence tending to show that a post-Cretaceous quartz-monzonite batholith underlies a large part of central Colorado, and that this batholith may have furnished the solutions from which was deposited ore in several districts. The eight widely separated exposures of quartz monzonite and closely related quartz diorite shown on the map (fig 12) are very likely the tops of masses connected below by a single batholith of which they are part. These upward extensions of the probable batholith have reached a higher level in the process of intrusion, and form roughly domelike masses, called *cupolas*, on the main batholith which has far greater horizontal extent at a lower level. There is no reason to doubt that other cupolas on the same batholith exist, but have not yet been uncovered by erosion.

Phenomena observed in Buckskin Gulch northwest of Alma, and described in the paper cited, indicate that the quartz monzonite was intruded after the early porphyries, and hence rose much higher than the roots of the porphyry dikes and sheets. This condition could, and probably did, result in bringing large volumes of magmatic solutions to a comparatively high level where leakage from a slowly solidifying large body was continuous through a long period. This hypothesis implies that, although the ores and porphyry originally came from the same reservoir, the reservoir with irregularly shaped top slowly worked its way upward subsequent to the rapid porphyry injection, and that the high parts of the batholith which fed ore minerals to any locality may or may not be far from a former vent through which porphyry magma flowed to the same locality. The reader may see in figure 12 that Red Cliff is about 12 miles from the nearest

[22]Crawford, R. D., A contribution to the igneous geology of central Colorado: Am. Jour. of Science, vol. 7, 1924, pp. 365-388.

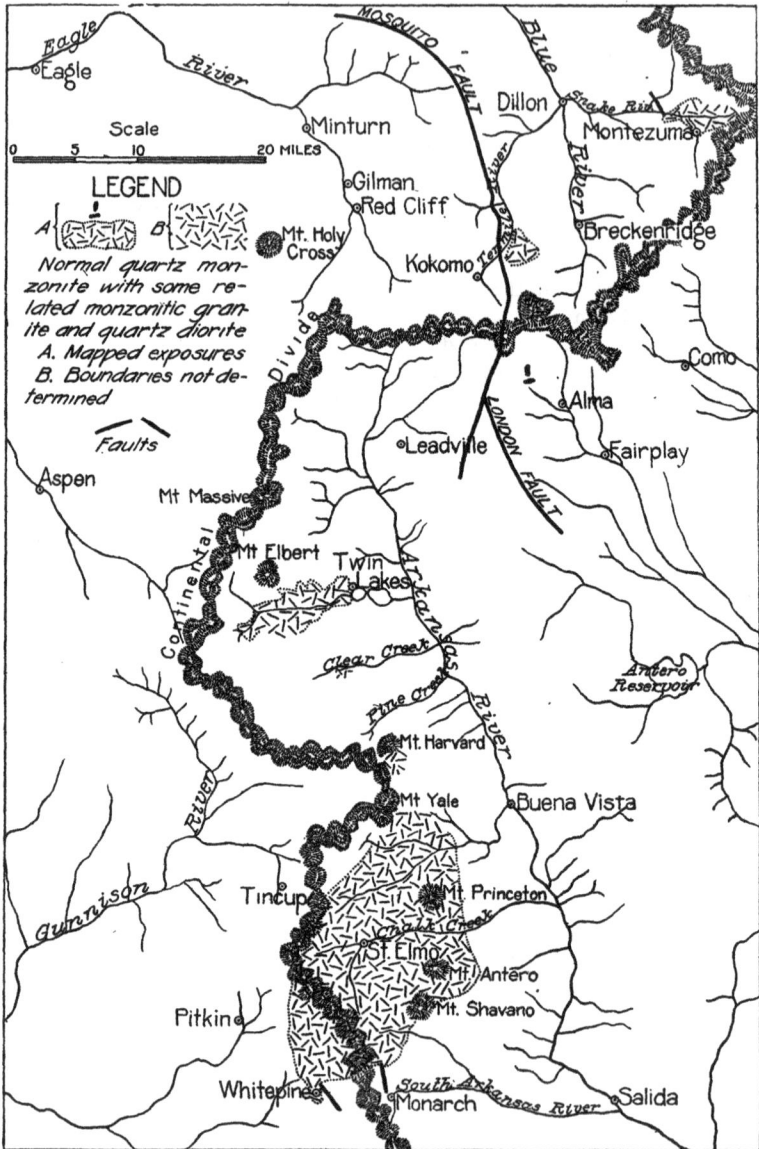

Figure 12. Map showing positions of post-Paleozoic quartz monzonite and related quartz diorite.

discovered post-Cretaceous quartz monzonite. Though this body may have a considerable underground extension to the west, structural conditions do not favor this particular stock as the source of the ores of this district. It is not improbable that the Red Cliff district is underlain by one or more cupolas of the quartz-monzonite batholith, that from and through the cupolas have flowed large volumes of metal-bearing magmatic solutions, and that these solutions effected ore deposition as well as extensive replacement by siderite, dolomite, and jasperoid.

FUTURE POSSIBILITIES

How far the ore shoots of Battle Mountain may be expected to continue to the northeast depends largely on the direction of flow of the depositing solutions. If the faulting resulted from igneous intrusion directly below and solutions rose vertically—conditions suggested by the sequence of metals—the extent of ore bodies toward the northeast should be limited by the size of the intrusion. If the solutions came from the northeast—a source unproved, but indicated by structure—the ores may be expected to extend far beyond the present mine workings; a dip of 10° is too small to lead to an unfavorable change in conditions of temperature and pressure within a short distance. Whether or not the largest shoots extend much farther northeast than the present workings there is no apparent reason why other ore bodies along known faults should not be discovered at the same depth as those now being mined.

The jasperoid and coarse dolomite previously described point to replacement in and near earlier faults. In several places faults are found near the dolomite bodies, and many faults have been preserved in the jasperoid. Jasperoid is very common in the Leadville district where it is not infrequently found near the oxidized ore bodies and, in places, grading into ore. Emmons[23] states that in the Tenmile district jasperoid occurs on the outer edge of ore shoots, and forms a transition zone between ore and unaltered limestone. In the Red Cliff district jasperoid was found near the surface in several mines and prospects on Battle Mountain. Ore occurs near jasperoid and coarse dolomite in several small mines near the mouth of Turkey Creek. A pocket of pyrite in jasperoid was opened by a prospect on Coal Creek. Oxides of iron and manganese are common in the faulted quartzite near jasperoid bodies south of Elk Creek and on both sides of Eagle River between Pando and Red Cliff. It is evident that both jasperoid and dolomite were deposited after—though in part perhaps contemporaneously with—the ore minerals, and it is probable that solutions that brought in both metallic and non-metallic minerals had a common source.

Unless the mineral-bearing solutions came from the southwest—the most improbable direction—good prospecting ground should lie east or northeast of any jasperoid or coarse dolomite body in which or under which a fault may be traced. Favorable localities, in the writer's opinion, are about three-fourths of a mile north of Gilman, near Turkey Creek, between Silver Creek and Coal Creek, and on both sides of Elk Creek. In several of these places prospectors have noticed the metallic mineralization, and

[23]U. S. Geol. Survey, Folio 48.

some sinking and tunneling have been done.   Ore bodies have been discovered at shallow depth on Turkey Creek, but in most or all of the other
places workings have not gone through the jasperoid.   To be of greatest
value work should be done at or beyond the northeast border of the
jasperoid, preferably by crosscutting the northeast-striking faults.   This
position, in part of the localities named, can be determined only by extensive prospecting.   If the solutions came from the northeast the chances
for finding large bodies of sulphide would seem to be better than if the
solutions came from any other direction.   Even then the ratio between
possible metallic sulphides and non-metallic minerals is probably very small,
and there may be faults entirely free from sulphides.

In Chapter V mention was made of faults in the quartzite, traceable
only short distances.   Between Coal Creek and Turkey Creek are several
of these faults and fault zones in which erosion has cut to considerable
depth; some of these are covered by soil and grass or timber.   Where
erosion has cut to greater depth the faults are covered by landslides.
Whether erosion cut deep into rock weakened only by faulting or by
faulting and subsequent mineralization only prospecting can determine.
Several of these places may be seen east of Eagle River between Turkey
and Coal creeks.   None has been seriously prospected, though many shallow prospects are seen in the quartzite farther west where considerable
iron and manganese are found.   To reach the Leadville limestone under or
east of the loose rock covering would entail less expense than to drill
through the large bodies of jasperoid.

CHAPTER VII
DESCRIPTIONS OF MINES

In this chapter no attempt is made to describe fully all the mines of the district. Many are closed or caved; some are filled or partly filled with water; and the known ore bodies of others have been nearly or quite exhausted. A fairly thorough study was made of the geology of the Mabel mine in the granite, the Ground Hog mine in the Sawatch quartzite, and the newer workings of the Eagle mines in the Leadville limestone. The geology of these mines well represents the geology of the productive area, and the three are described below in considerable detail. Many mines in the oxide zone are briefly described, but some of the largest former producers are passed with scant notice and without description. Among these—part in quartzite and part in the limestone—are the Black Iron, Cleveland, Belden, Iron Mask, Little Chief, Bleak House, and Rocky Point mines. The old workings are still accessible in many places, and the great size of some of the former ore bodies is made evident by stopes and drifts still open. Because the Ground Hog mine is the most easily accessible of all the larger mines in the quartzite a detailed study of structure and ore occurrence was made in the Ground Hog, though larger ore bodies may have been found in the quartzite in other mines.

In the descriptions that follow the term granite is used for both granite and related quartz monzonite, and the term limestone is used for the calcareous rock of the Leadville formation which is in part true dolomite.

MABEL MINE

BY RUSSELL GIBSON

The Mabel mine was opened by B. A. Hart in 1900, and was still producing ore in the summer of 1923. It is one of the largest mines in the district in the granite, and has, according to Mr. Hart, produced $325,000 worth of ore which has averaged 2 to 4 ounces gold and 5 to 15 ounces silver per ton. In addition to gold and silver, the ore carries 8 to 10 per cent lead, and copper up to 3 per cent. One shoot is reported to have produced $40,000 worth of ore that averaged $275 per ton.

The Mabel mine has 3,600 feet of development on five levels which are connected by an inclined shaft 264 feet deep on the pitch. (See figs. 13 and 14.) The shaft follows the dip of the Mabel vein which increases from 56° at the collar of the shaft to 66° below the second level. Most of the ore comes from two fissure veins: the Mable vein which strikes N. 30° to 40° E., and a cross vein striking N. 55° to 65° E. In general the veins dip southeast. Observations on the first level indicate that the Mabel vein occupies a fissure along which there was faulting prior to ore deposition. The fault is normal, the hanging wall apparently having moved down a distance which could not be determined. The cross vein, which strikes N. 55° to 65° E., is intersected 180 feet northeast of the shaft on the first level, and at shorter distances in the same direction from the shaft on each succeeding lower level. Like the Mabel this vein increases

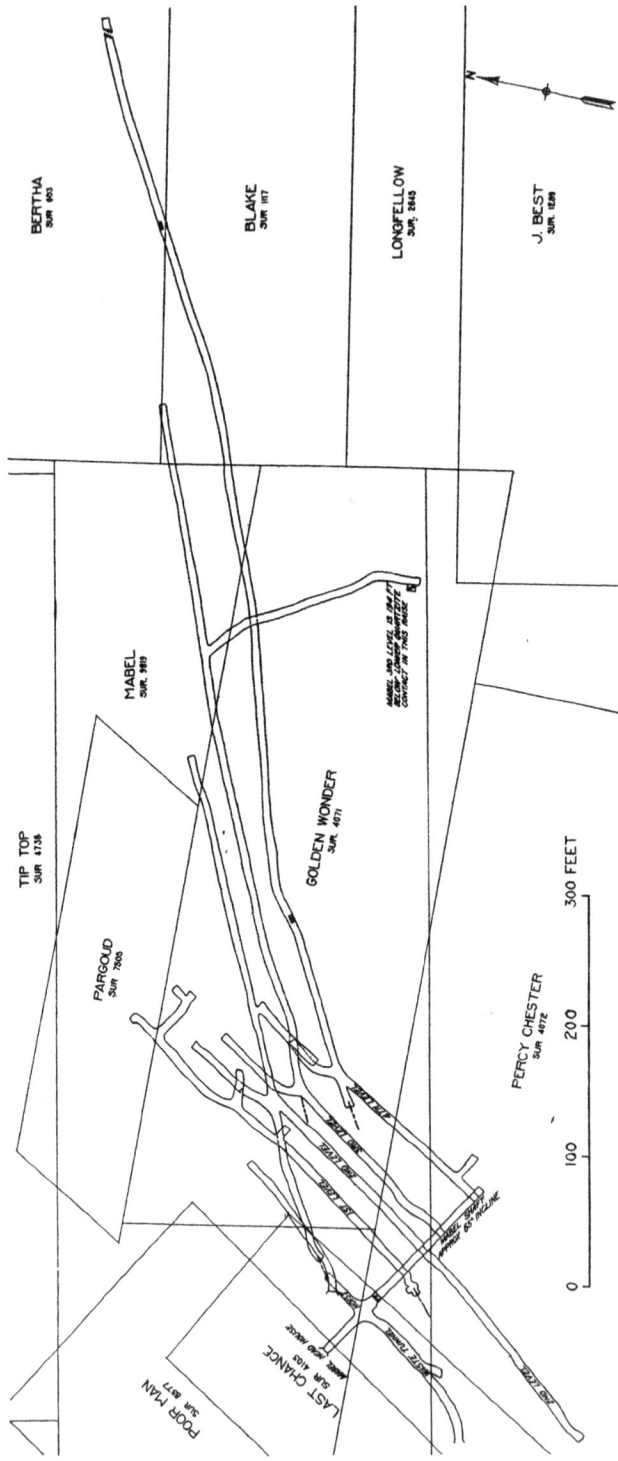

Figure 13. Plan of the Metal mine, reduced from map furnished by J. M. Dismant

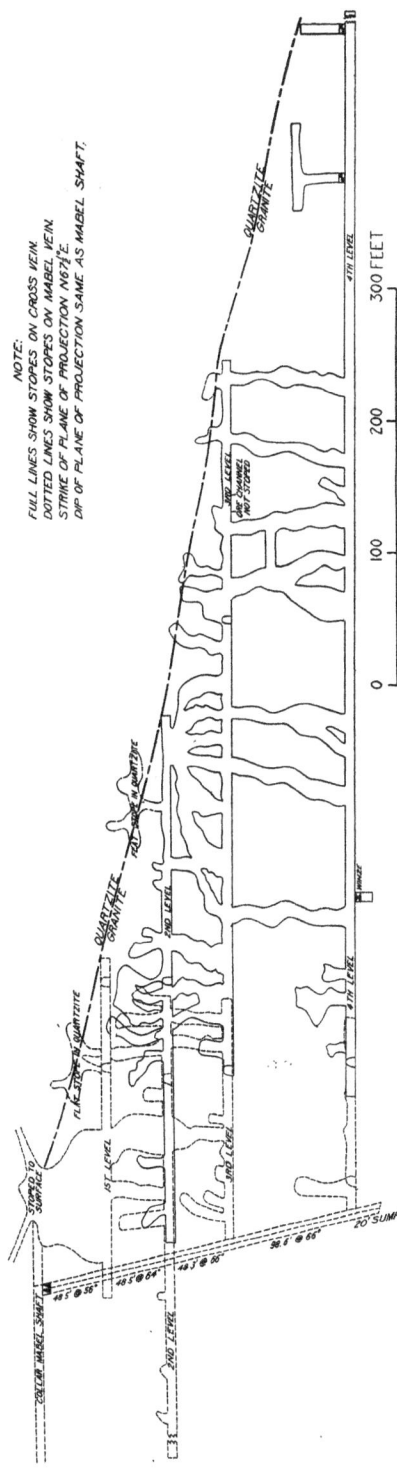

NOTE:
FULL LINES SHOW STOPES ON CROSS VEIN.
DOTTED LINES SHOW STOPES ON MABEL VEIN.
STRIKE OF PLANE OF PROJECTION N67½°E.
DIP OF PLANE OF PROJECTION SAME AS MABEL SHAFT.

Figure 14. Section through the Mabel vein, reduced from drawing furnished
by J. M. Dismant

in dip with depth, from 56° on the first level to 76° on the lowest drift. Where the Mabel vein intersects the cross vein on the first and fourth levels relationships indicate that the Mabel is the older. Elsewhere the evidence is more obscure. But in these two places the Mabel vein is offset by the cross vein, and hence may have been mineralized before the movement took place along the cross fissure.

The amount and character of the mineralization in the two veins is similar. The chief minerals are pyrite, copper-bearing pyrite, sphalerite, galena, and quartz. The order of deposition, so far as could be determined from remnants of ore, is not regular; but pyrite, the most abundant mineral, seems to be one of the first minerals deposited and quartz one of the last. These minerals occur in streaks from less than 1 inch to 10 inches thick, and less commonly in vugs. The wall rock is sparingly replaced close to the vein.

In most places the fissured zone varies in width from 2 to 6 feet, and is made up of parallel cracks, not all of which are filled with sulphide. Commonly one thick vein splits into two veins or fingers into several, some of which coalesce farther on. The associated decomposed wall rock and gouge contain more or less disseminated pyrite in good cubes and pyritohedrons. According to Mr. Hart, richer ore was found near the intersections of veins. The fissures are tighter and the veins grow thinner and leaner with depth, especially toward the northeast extremities of the lower levels. Here the stopes are smaller and the average length of the stulls does not exceed 2 feet, whereas elsewhere many stopes are connected through several levels, and are in places 8 feet wide. Where the drifts intersect the lower Sawatch quartzite contact the veins run up into the quartzite but thin rapidly at the contact, in places almost disappearing. The contact, however, and the bedding planes above show good mineralization; and the quartzite is partly replaced by pyrite.

The mine is equipped with a steam hoist and skip, and electric lights. An aerial tram connects the head-house with the Mabel-Pursey Chester mines switch.

GROUND HOG MINE

By E. A. HALL

The workings of the Ground Hog mine extend down the dip of the quartzite in a general direction of N. 45° E. for a distance of 1,500 feet. (See fig 15.) The dip of the quartzite is 12° to 15° N. 40° to 45° E., and as the inclines go into the side of Battle Mountain the deepest workings are about 800 feet below the surface. The workings extend about 1,400 feet in width in a northwest direction. The mine has been worked from twelve main inclines and several minor ones. At the present time not all of the inclines open at the surface. However, they are connected underground. Once inside the mine it is possible to go through the entire workings, with the exception of a few drifts filled with water or waste and a few minor drifts from the surface not joined with the main workings. The mine was also formerly connected with the surface by several shafts, all of which have fallen into disuse and have become partly filled or destroyed. In many places the side drifts become very small, usually about 2.5 to 3 feet high, with scarcely enough room for a man to crawl through, though they extend several hundred feet in length. The main inclines and crosscuts are sufficiently high to allow a man to walk erect, except in a few places where the quartzite is particularly hard and where tracks have taken up some of the space.

At the present time five of the inclines and the Nottingham crosscut are tracked and others are partly tracked; only two inclines, the Doddridge and Nottingham, and the main crosscut are used for haulage purposes. There are two head-houses in repair at the openings of the Nottingham and Doddridge inclines, and bear their respective names. Mine and buildings are not wired for electricity, hence gasoline and steam are used for hoisting. At one time a considerable part of the mine was piped so that air could be used for drilling, but the air system was out of repair at the time of our examination.

PRODUCTION

Although a great part of the production of the Ground Hog mine was from rich shoots or pockets for which records of approximate production have been kept, it is next to impossible to make an estimate of the total production of the mine. The Forgy "bullion hole," located in the Nottingham (right) incline about 225 feet above the main crosscut, produced about $37,000 worth of ore. The Doddridge winze, in the Nottingham incline 100 feet below the main crosscut, produced about $50,000 worth of ore. A small stope on No. 4 South drift produced about $4,000 worth of ore carrying $15.00 in gold and 15 ounces of silver per ton. No. 5 South produced about $47,000 worth of ore. The Baldwin and Oneal "bullion place," 300 feet down the Doddridge incline, produced $12,000 worth of oxide ore. The Woods and Henderson "bullion shoot," 200 feet down the same incline, produced $10,000 in oxidized ore. The total production in the Doddridge incline area below the main crosscut is said to have been about $500,000.

The above figures were supplied by Dismant Brothers of Red Cliff, and represent the production of only a few of the richest shoots or "bullion" pockets in the mine. However, from these figures ($660,000) it can be seen that the mine produced much ore.

FAULTING

The fissures and faults apparently trend in two general directions. Those of the first group strike about N. 35° W., and compose the minor system. Those of the second group strike N. 40° to 45° E., and compose the major system.

The breaks of the minor group are nearly vertical, and each forms a single sharp and almost straight fissure. None of the northwest-striking fissures shows evidence of faulting, but all seem to have originated from a common cause. These fissures occur very regularly, and often not more than 10 feet apart. Hence a considerable amount of displacement or adjustment could have taken place, and yet when distributed among so many breaks the chances are that there would be no perceptible evidence of movement in any single fissure. This group of fissures very closely resembles jointing in igneous rock.

The northeast-striking faults of the major system are for the most part about vertical. These are not single fractures, but are grouped in zones of faulting, each zone being composed of several fractures. Many of the different faults in a zone come in and leave the zone at a small angle. In practically all of the strong northeast-striking faults slicken-siding, grooving, displacement, or other evidence of faulting can be seen.

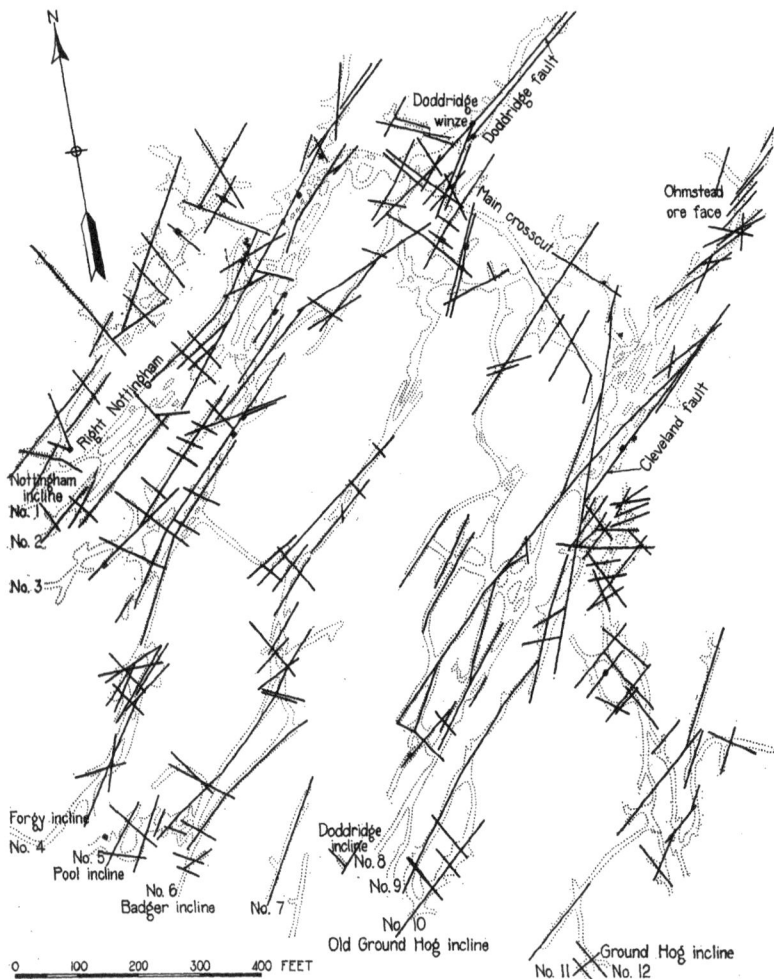

Figure 15. Map showing faults and fissures in the Ground Hog mine; geology mapped by E. A. Hall

Probably the two most important faults in the mine are the Cleveland and the Doddridge faults.

Cleveland fault.—This fault strikes N. 45° to 55° E., and is vertical. It can best be seen about 40 feet east of the Doddridge incline opposite the main crosscut. The faulted zone is 12 to 15 feet wide, and contains immense quantities of gouge and breccia. One minor fracture shows a displacement of 3 feet, and the fault surface is well grooved and slickensided. The grooves are vertical, showing that the blocks moved up or down with relation to each other. On the surface small separated faults probably belonging to the Cleveland zone indicate very little actual displacement, and give no evidence of great movement. It is possible that the zone represents considerable movement back and forth. This fault contains red iron oxide, brown iron oxide, and many quartz crystals.

Doddridge fault.—This fault can best be seen at the Doddridge winze 100 feet below the main crosscut in the Nottingham incline. It strikes N. 45° E., and is vertical. The zone is about 10 feet wide, and is well brecciated and filled with gouge. The Doddridge winze has been sunk 160 feet on the fault to the granite, but, owing to water, only two levels below could be examined. On the two lower levels very highly polished mirror-like slickensiding was noticed. The grooves are horizontal, and the movement must have been at right angles to the direction of movement of the Cleveland fault. The fault zone is partly filled with sulphide ore and quartz gangue. In all probability the deposition of most of the ore in the mine was controlled by the northeast-striking faults.

Bedding faults.—In addition to the faults described many bedding faults of various sizes were encountered in the mine. They were often cut in two or displaced by other faults. As a rule these faults are tightly filled with gouge between walls 2 to 4 inches apart, and were not found to be mineralized.

CHARACTER OF WALL ROCK

The bed which has been followed by the greater part of the Ground Hog workings is a medium-grained porous sandstone or quartzite which dips 10° to 12° northeast. It is not as well silicified as most of the quartzite formation, and when not entirely replaced by ore the bed is dark colored as if it had been stained with iron-bearing solutions. This porous, non-resistant quartzite bed averages about 2 feet in thickness. Below it is the usual white, pure, compact, fine-grained quartzite. Above the porous bed is quartzite very little different from the quartzite below the bed, excepting that it appears to be even more dense and compact, whiter, and purer than the usual quartzite. In fact the upper quartzite bed is so dense and hard that it shatters like glass or pure quartz when struck with a hammer. Stopes Nos. 3 and 4 South have been driven up into the quartzite a distance of 35 to 40 feet. About 30 feet above the main level the usual compact, fine-grained quartzite seemingly grades into more shaly members. Some of these shaly beds are calcareous, and were readily replaced by ore.

ORE DEPOSITS

The ore deposits of the Ground Hog mine are of two types closely associated with each other. These are fissure deposits and replacement deposits.

Fissure deposits.—Nearly every fault, fissure, and fracture in the quartzite in the Ground Hog mine is partly filled with mineral. However, the fissures of the minor system are very narrow, and allow little room for deposition from solution. Thus the veins striking northwest are of very little economic importance except where they intersect a vein of the other system. The major faults striking northeast seem to have controlled the greater part of the ore-bearing solutions. In addition to being wider than the fissures striking northwest, they are more closely packed with mineral. Moreover, the solutions effected a variable amount of replacement of the quartzite—usually several inches— on both sides of the faults of the major group. The fissure deposits are characterized by a very pronounced crustification. Crystals of sphalerite were deposited over the pyritohedrons of pyrite; the last deposition was of quartz or barite. Some of the highest-grade ore of the mine occurs as fissure deposits in the faults and fractures striking northeast. Invariably, however, the fissure deposits pinch out within a few feet above and below the main level.

Replacement deposits.—As was stated in connection with the fissure deposits, the ore-bearing solutions effected the replacement of several inches of wall rock on each side of the fissure. In many places the mineral was found disseminated through the quartzite beyond the zone or limits of complete replacement. However, the most important replacement deposit in the mine was the almost complete replacement, especially near one of the faults or fissures, of the porous, non-resistant two-foot bed of sandy quartzite. This particular bed must have given to the ore solutions a channel of easy circulation. It was this easily replaceable bed that made the mine a commercial possibility: first, because the bed was much easier to mine than the hard quartzite, thus reducing the cost of mining; second, because the bed itself contained a large amount of ore, especially where entirely replaced.

The ore is of two general types: the first is compact and hard; the second is fragmental and granular. Though the second retains the structure of the original quartzite it is not as well cemented as the first. Ordinarily the granular ore has been of lower grade than the other, and though it is easier to mine it has been looked upon with disfavor by the lessees. At the time of examination the character of the replacement could be seen at the Ohmstead ore face about 150 feet from the lower end of the Doddridge incline. At this face the replaced portion varied from 6 inches to 2 feet thick. The replacing material (pyrite) carried a small amount of gold and about 50 ounces of silver per ton.

The mineralized zone in and near the Doddridge fault was probably a combination of fissure and replacement deposits. This zone has been stoped out 35 feet overhead. The Doddridge winze, which sinks 160 feet to the granite, was examined as far as the water would allow. For a distance of about 80 feet below the main level the rock of the fault zone was replaced by or filled with solid pyrite varying from 5 to 10 feet in thickness and extending laterally 40 to 50 feet. This body of ore is too low in grade to work.

Character of the ore.—Oxidized ore extends down the dip of the quartzite for a distance of 500 to 700 feet. This zone included some of the richest "bullion" pockets, but now no valuable ore bodies are known in the oxidized zone. The oxides were chiefly those of iron and a small amount of manganese. The manganese oxide seems to have been limited to a few of the largest northeast-striking faults, such as the Cleveland fault.

The sulphide ores below the oxidized ores continue down the dip of the quartzite to the deepest faces in the mine. The sulphides are galena, sphalerite, chalcocite, copper-bearing pyrite, and pyrite. The last is by far the most important, while the others occur sparingly in the small area near Nos. 3 and 4 South drifts and stopes. In the sulphide zone pyrite crystals (pyritohedrons) are very common. The pyritohedrons are well shaped and uncombined with other forms. Some are 2 inches in diameter. Small cubes and octahedrons of pyrite were found on the lower levels of the Doddridge winze area. Most of the pyrite carries silver; some of it runs as high as 60 ounces per ton. Galena also carries silver, up to 50 ounces to the ton. The silver probably occurs in the form of argentite, silver sulphide. The oxidized ore carries very little silver.

Secondary-sulphide enrichment is indicated by the presence of small quantities of chalcocite, covellite, and probably bornite in the stoped area of Nos. 2, 3, and 4 South drifts. Here thin coatings of covellite were found on crystals of pyrite, in addition to granular to massive chalcocite associated with pyrite.

Native gold has been found in both the oxide and sulphide zones, but samples collected in 1921 show only small quantities of gold in the sulphide ore. Much of the gold, particularly in the oxide zone, was in nuggets; and a large quantity of high-grade gold ore has been produced by this mine.

EAGLE MINES

By R. D. CRAWFORD

In recent years the Empire Zinc Company has developed a group of claims east, northeast, and southeast of Gilman, and has produced from these claims a very large tonnage of sulphide ore. The property of this company includes also the Bleak House, Wilkesbarre, Little Chief, Iron Mask, Rocky Point, Belden, F. C. Garbutt, Black Iron, and others; some of these mines have been producers from the early eighties. These mines, formerly worked by several companies, were bought in 1915 by the Empire Zinc Company which has worked the group continuously since the purchase. An average force of about 175 is employed. Under the present ownership the entire group is called the Eagle Mines.

The newer workings are entered through the Wilkesbarre shaft (Eagle No. 1) north of Gilman and through the old Black Iron tunnel and incline (Eagle No. 2) at Bell's Camp. Most of the ore is hauled out through the Newhouse tunnel whose portal is near the bottom of the canyon at Belden. The old mine workings, including adits, inclines, and stopes, are still accessible in many places. One may get an idea of the size of the developed area of the Eagle mines from the following statements: The Newhouse tunnel portal is S. 21° W. of the Wilkesbarre shaft and distant about 2,300 feet. The Black Iron tunnel portal is S. 17° E. of the Wilkesbarre shaft and distant about 5,760 feet, or more than a mile. The shortest underground connection between the last two points is more than 6,000 feet long. Six sulphide-ore bodies have been proved in the newer workings, and two of them have yielded an unusually large tonnage.

The writer traversed all the deeper workings of the Eagle mines for the purpose of examining and platting faults and other structural features that might have a bearing on the occurrence of ore. The mineral character of the ores and many details of ore structure were noted, but no systematic sampling or thorough study of the ores was made. Many of the details given here were generously furnished by Mr. A. H. Buck of the Empire Zinc Company.

Part of the old workings of the Little Chief, Iron Mask, Belden, and Black Iron were hastily examined, but nearly all the time spent in the mines was employed in the stopes and on the lower levels given below with average or approximate elevations:

Fourth level at 8,830 feet, Eagle No. 2
Fifth level at 8,805 feet, Eagle No. 2
Sixth level at 8,735 feet, Eagle No. 2
Seventh level at 8,770 feet, off Belden incline
Fourteenth level at 8,660 feet, connects Eagle No. 1 and No. 2
Fifteenth level at 8,560 feet, Eagle No. 1
Sixteenth level, including Newhouse tunnel, at 8,505 feet

GEOLOGIC FEATURES

Though much ore has been produced from the Sawatch quartzite by the Bleak House and Rocky Point mines, by far the greatest amount produced by the group in recent years has come from the upper part of the Leadville limestone. The exceptional "chimney shoot" of Eagle No. 2 was nearly solid sulphide through a vertical range of 200 feet, thus extending from top to bottom of the Leadville formation. The quartzite member of this formation underwent about the same degree of replacement by this ore as did the overlying and underlying limestone.

All the ore examined lies in or near rock broken by faulting. Two principal kinds of faulting have affected the rock, and their results may be seen in barren ground between or below the ore bodies. The first type was differential movement along planes nearly or quite parallel to the bedding planes of the rocks; in the second type the movement was along planes that cut across the beds at high angles and are nearly vertical. The majority of these faults of high dip strike northeastward and are thus dip faults,—that is, they strike in the direction of dip of the beds. At least two important faults of high dip strike in other directions: one, on the 6th level near No. 2 shaft, strikes about N. 80° E.; the other, on the 16th level near the extreme east workings, strikes nearly N. 45° W. The age of the last mentioned fault has not been determined. It has been considered possibly younger than the ore, but the writer could see no sign of post-mineral faulting in the ore face in the stope. The principal northeast-striking faults are older than the ore. In addition to the bedding faults and those of high dip are a few unimportant faults and breaks of intermediate dip and random strike.

In barren ground at several places in the mines are narrow fissures and veins whose walls show no indication of having slipped. Most of these fissures are not more than 2 or 3 inches wide. Some are open and dry; some are partly filled with running water; others are partly or wholly filled with dolomite or pyrite.

Caves a few inches to several feet wide are found at the border of the ore bodies or are formed in the process of mining. The caves and pockets contain dolomite sand which runs out in large quantity when the cave is tapped. Though the sand is composed chiefly of rounded and angular grains of dolomite some quartz is present. The stratification sometimes seen and the rounded character of the grains show that much of the sand has been deposited in water courses. However, a large part of the so-called sand on the borders of the stopes is disintegrated dolomite from the walls. The coarse crystalline dolomite slacks and becomes friable in the mine workings and probably also on the walls of the caves.

In the Rocky Point mine, not examined in detail by the field party, gold-bearing pyrite was found in the upper half of the Sawatch quartzite about 150 feet below the Leadville limestone. In the Leadville limestone, just below the quartzite member, was a copper-ore shoot 200 feet long and 30 to 50 feet wide. The ore here was chalcopyrite or copper-bearing pyrite. Above the quartzite member of the limestone and nearer to the quartzite than to the overlying shale was the No. 1 zinc-ore shoot described below. All three of these shoots were in the same vertical plane, the last two in the Iron Mask claim.

SIZE AND RANGE OF ORE BODIES

All the ore bodies of the Eagle mines have their longest dimensions in the direction of dip of the inclosing rocks.

No. 1 ore shoot has been developed in the newer workings from the 11th level to the 16th level through a vertical range of 200 feet, and has been proved by drilling to a depth of 200 feet below the 16th level. The horizontal range of the shoot in the newer workings, including the ore proved by drill holes, is 2,000 feet. Through 700 feet of the distance in the deeper part of the mine the shoot is about 100 feet wide. Cross sections of the shoot are roughly elliptical, the ellipses having very irregular peripheries. The long axes of the ellipses are nearly horizontal and range from 75 to 150 feet in length; the short axes are 50 to 100 feet long. All the ore here is sulphide. Before the mines were acquired by the Empire Zinc Company a large part of this shoot had been mined through the Iron Mask, Little Chief, and other inclines. Most of the earlier mined ore was oxidized or carbonate ore. The vertical range of oxidized and sulphide ore in this shoot, as far as it has been proved, is 630 feet. The pitch length of the proved shoot is 3,060 feet. The ore has been continuous from the surface outcrop to the greatest known depth of the shoot.

No. 2 ore body differs from No. 1 ore body in that it is a blanket vein (Pl. III, in pocket). Excepting that part known as the chimney shoot the thickness ranges from 6 to 50 feet, and averages about 20 feet. The ore body is from 75 to 150 feet wide. The "chimney" is part of the No. 2 shoot, and extends below the blanket roughly in the shape of an inverted cone. The axes of the elliptical base of the cone (the upper part) are 90 and 220 feet respectively. The height of the "chimney", including the blanket, is 200 feet. No. 2 shoot, in the newer workings and in the old Black Iron mine, has a proved pitch length of 2,880 feet. The ore has been continuous throughout this length from the surface. The oxidized zone extended 1,200 feet from the surface down the pitch. About two-thirds of the way down to the deepest workings is the center of the chimney shoot described above.

No. 3 ore shoot is evidently the same as the largest one of the Belden mine worked many years through the Belden incline. Excepting about 200 feet of unexplored ground between the 7th level of the Belden and the new 14th level, the shoot has been developed through a length of 1,400 feet. The lowest stopes are on the 14th level where the width is nearly 100 feet.

No. 4 shoot has been mined from the surface through a length of more than 600 feet, and has been proved through an additional 150 feet. Its average width is about 40 feet. Other ore bodies have been partly developed, but their extent is imperfectly known.

CHARACTER OF THE ORES

The metals produced in greatest quantity from the sulphide-ore bodies are zinc and silver. Part of the ore carries a very little gold, part carries copper, and some lead is produced from galena found in only a few places. The two most abundant ore minerals are zinc blende and pyrite. The lead content of the blende varies from less than 1 per cent to 15 per cent. The blende ordinarily carries no silver, but in places small quantities have been found carrying silver up to 12 ounces per ton. The lead content in different ore bodies varies from 18 per cent to less than .5 per cent. Different ore shoots show different relations between lead and zinc sulphide. In some shoots the lead is found in larger proportion just below the oxidized zone, and the lead content decreases with depth; in other shoots the reverse is true. Throughout the mine the gold content is nearly negligible.

Practically all the silver produced is found in pyrite. The galena does not carry silver in paying quantity. Silver values may be very spotted in pyrite in a single stope; within a small area assays may run from 5 to 150 ounces per ton, with an average of 10 to 15 ounces in carload lots. The copper content is very low and nearly negligible; small lots have assayed as high as 8 per cent copper. As a rule high silver accompanies high copper content, but the silver content does not necessarily rise in the same proportion as the copper. In a few places high copper is not accompanied by high silver. In places rich silver-bearing pyrite carries practically no copper.

MINERAL RELATIONSHIPS IN ORE BODIES

The relation between minerals is not uniform in all the ore bodies. In No. 2 ore shoot there is a large core of pyrite surrounded by a shell of zinc blende from 10 to 20 feet thick. The line of demarkation between the two is very distinct. In this shoot the line between high-grade zinc ore and wall rock is as sharp as that between pyrite and zinc ore. A similar sharp boundary between ore and wall rock is found on the 16th level (16 E.)

The zinc ore of No. 1 shoot is uniformly lower in grade than that of No. 2 shoot. It carries more lead, and has pyrite and siderite in varying but small quantities throughout the body. Between levels 15 and 16 there is a body of pyrite on the south side of the main body of zinc ore. It has no value as ore, and its extent is unknown.

Siderite is found in varying quantities near the ore bodies; in places the siderite masses are many feet thick. In the vicinity of lead-free zinc ore there is very little siderite, but where the lead content increases the amount of siderite surrounding the ore body also increases. Coarse crystalline dolomite is common in and near the fault zones under the ore and laterally. Though this dolomite locally forms the walls of ore bodies finely crystalline dolomitic limestone is the commonest wall rock. In places the contact between ore and wall rock is sharp; in fewer places there is a gradation from ore to wall rock. Shale, gouge, and broken quartzite and porphyry overlie the ore, and are sometimes found at the sides and within the ore.

MINING METHODS AND EQUIPMENT

The mines are supplied with electric power and lighted by electricity. All the drilling is done by Denver Rock Drill Company machine drills. Thus far only the square-set method is used, though the managers are planning to use a modification of the cut-and-fill method on one of the ore bodies. Ore is moved in the main haulage ways in trains of five-ton cars hauled by storage-battery locomotives. Tracks are laid at a two-foot gauge with thirty-pound rails on 6 by 8 inch spruce ties. The maximum grade is 1 per cent. On the 14th level there is about 1.5 miles of track for the locomotives, and on the 16th level about 1.25 miles. The ore is transferred from the 14th level to the 16th level through three ore pockets each 144 feet deep and 12 feet by 20 feet in cross section. The pockets are in solid rock and have no timbers in them. The capacity of each pocket is approximately 3,000 tons. From these pockets the ore is hauled on the 16th level to three loading pockets near the portal of the Newhouse tunnel. Each of the loading pockets has a capacity of about 500 tons. A short tunnel runs from the surface to the bottom of these pockets. On this lowest level an electric lorry car, having a capacity of 12.5 tons, hauls the ore from the loading pockets, and dumps it directly into the railroad cars on the siding at Belden.

The zinc ore is shipped to the Company's plant at Canon City, Colorado, and the silver ore is shipped to the American Smelting and Refining Company's smelter at Garfield, Utah.

The Empire Zinc Company maintains a club house and modern hospital at Gilman. A physician and trained nurse are in constant attendance at the hospital, though their services are not often required. The Company has a boarding house and a rooming house for single men. Married employees may rent from the Company attractive dwelling houses each having four rooms and bath. The Company owns 35 of these houses each with running water and connected with the sewer system. Home owners in Gilman are permitted to connect with the Company's sewer system by paying the cost of connection; there is no charge or tax for maintenance.

EVENING STAR MINE[24]

The Evening Star tunnel trends southwest 485 feet in the granite. The tunnel has been driven on a fissured zone where a narrow porphyry dike

---

[24]This and the descriptions that follow were written by Russell Gibson. Excepting the three previously described, the mines follow in the order of the list at the margin of the map (Pl. I).

has intruded the granite. This mine is said to have produced gold, silver, copper, and lead.

BEN BUTLER MINE

The Ben Butler mine has been worked through two inclines in the quartzite and a tunnel in the pre-Cambrian rocks below. The mine has 1,100 feet of development on the tunnel level. At least four fissure veins striking northeast were prospected. One of the largest is at the contact between granite and quartz diorite. The fissured zones vary in width from 1 to 7 feet, and contain streaks of pyrite which carries gold and silver; they also contain streaks of vein quartz. There is little replacement of the wall rock. Most of the ore on this level came from three fissures which were cut by the main tunnel between 660 and 720 feet from the portal. Some good ore was produced in a raise to the quartzite 440 feet from the portal of the tunnel.

The upper and lower inclines have a total of 2,500 feet of workings in the quartzite. They trend northeast, and follow two similarly mineralized beds where the ore is the result of replacement of quartzite by pyrite and, to a less extent, by other sulphides. Near the surface the ore has been oxidized. In places only a few inches of quartzite has been replaced; elsewhere there has been partial replacement to a thickness of 6 feet. Most of the fissures are small and show little movement; in most places they strike northeast. The mineralization in the fissures is thin and unimportant.

It is estimated that the total production of the mine has been about $250,000 worth of ore. The mine has not been worked for many years, but in 1923 Messrs. Henry and Frank Martin were cleaning out the tunnel for a haulage way for ore mined in the inclines.

TIP TOP TUNNEL

The Tip Top tunnel, which trends northeast for 1,400 feet in the pre-Cambrian rocks, reaches the quartzite at its face. The tunnel intersects 11 small fissures which are mineralized, some of which have been prospected by drifts and raises. With one exception these fissures strike northeast, and seemingly more or less movement has taken place along most of them. The vein material consists of pyrite, sphalerite, and quartz; the vein is rarely wide. Pyrite is disseminated through gouge and fault breccia which may be from one-half inch to 12 inches wide. The same mineral also replaces the granite wall rock to a slight extent a few inches from the fissure proper. A flat-lying fissure vein between the granite and quartzite shows an inch of sulphide, and the quartzite 6 inches from the contact contains disseminated pyrite.

STAR OF THE WEST INCLINE

The Star of the West is opened by an incline which trends northeast in the quartzite. The chief ore mineral is pyrite, and it occurs as replacement bodies in the quartzite. A few tight, scantily mineralized fissures were seen which are, in most places, approximately vertical and strike northeast. The mine has not been worked for some time, and the production is not known to the writer.

PURSEY CHESTER MINE

Mr. B. A. Hart, one of the owners of the Pursey Chester, has kindly furnished much of the following information concerning the early history, tenor of the ore, and production of the mine. The mine was opened in 1883 by A. H. Fulford and Judge Ackley. In 1885 J. T. Hart bought the property. The oxidized ores, which were produced in the beginning and which ran higher in gold than in silver, averaged $150 per ton. Rarely ore worth $2,000 per ton was produced. Later the sulphide zone was reached where silver was more abundant than gold. Returns have always been made for 4 to 5 per cent copper, but generally not for lead. In 1923 the ore averaged .41 ounces gold and 12 ounces silver per ton. Mr. Hart states that the total gross value of the ore produced to date is a million dollars.

The mine is opened by an incline which trends northeast in the quartzite. The wall rock shows numerous fissures almost all of which strike northeast; and, although they are fairly tight and only thinly mineralized, it is observed that drifts follow them, and that ore has been produced near them. Almost all the ore is the result of replacement of two beds or small groups of beds of quartzite which are separated by barren rock. In places these beds are partly replaced to a thickness of 4 or 5 feet, though the average is probably less than 4 feet. The chief mineral is pyrite; much of it is copper-bearing. Small amounts of chalcopyrite and sphalerite were seen by the writer. There has been little oxidation in the deeper workings of the mine, but nearer the portal of the incline remnants of limonite and altered pyrite may be seen. The stopes are not high, but are very wide and long; they pitch with the dip of the beds. The great size of the stopes indicates that thousands of tons of ore have been shipped.

The mine is equipped with an electric hoist and electric lights. The ore is trammed out through the Mabel whence it is delivered to the Mabel-Pursey Chester mines switch by an aerial tram.

POTVIN MINE AND F. C. GARBUTT MINE

These mines are opened by inclines which trend northeast, and seemingly follow a fault zone in the limestone. The workings converge and break through in several places. Although there is evidence of much disturbance in both mines, the fault or faults can not be described because many of the crosscuts and stopes are badly caved. In places the limestone is broken and brecciated from floor to back. The ore was evidently in pockety replacement bodies of irregular dimensions and in fissures. Remnants of the ore indicate that oxidation has been pretty thorough. At the time of the examination 900 feet of underground workings could be seen in the Garbutt mine and 1,000 feet in the Potvin.

In 1922 a shaft was sunk to open the Potvin incline, but no ore was shipped. In the summer of 1923 work had been discontinued.

ALPINE MINE

The Alpine has about 700 feet of development along fissures in the granite. The fissures strike northeast, dip steeply or are vertical, and contain as much as a few inches of pyrite. There has been a little replacement of the granite wall rock immediately adjacent to the fissures. The largest stope is about 100 feet long and 25 feet high.

A good blacksmith shop houses an air compressor, and a small mill has been erected.

PINE MARTIN MINE

The Pine Martin has at least five openings in the quartzite and thousands of feet of development, but the mine has not been worked for several years and is partly filled with water.

The ore occurs as irregular replacement bodies in the quartzite. In the lower workings the chief mineral is pyrite; nearer the surface are large masses of brown, earthy, limonitic material which, according to Mr. Smith who was in charge of the property in 1921, carries both gold and silver, but is low grade and difficult to treat. Some of the pyrite is cupriferous; manganese oxides are found in some places with the limonite. Two groups of thin, closely spaced, vertical fissures containing scant mineralization, were seen. One group strikes northeast, the other northwest.

According to the reports of the Director of the Mint the Pine Martin produced ore to the value of $204,339 in the years 1887-1890 and 1892 (gold and silver taken at coinage value).

The mine is equipped with electric lights. A mill building is connected with the main incline by a snow shed.

CHAMPION MINE

The Champion is opened by an incline in the quartzite which trends N. 51° E. When the mine was examined in 1923 over half of the workings were under water and could not be seen. The mine has not been operated since 1916.

Most of the ore was found in beds of quartzite more easily replaced than beds above and below; fissure filling is of minor importance. The ore bodies were very irregular in size and shape. Ore was cleaned out of pockets only 2 inches across in some parts of the mine; elsewhere there are wide stopes 5 feet high. The chief minerals are pyrite (in part cupriferous) and iron oxides; small amounts of siderite and manganese oxides were seen. The fissures in the accessible workings strike northeast or northwest, and are nearly vertical.

The total production of the Champion is not known to the writer. According to the Mint reports the mine produced gold and silver having a coinage value of $100,859 during the years 1887-1892.

The mine is equipped with electric lights, a steam hoist, and an electrically driven air compressor.

BODY MINE

The Body mine is developed by two tunnels driven on fault fissures in the granite. Most of the ore has come from one fissure vein which strikes southwest and dips southeast or is vertical. The upper tunnel is 495 feet long, and the lower, from which little ore was taken, is 460 feet. Two stopes in the upper tunnel are approximately 35 feet high, and a third stope is 20 feet high. The fissured zone is rarely 36 inches wide; the maximum width of mineralization is 18 inches. The ore seen consists chiefly of sphalerite, pyrite, and galena, in streaks from a fraction of an inch to 3 inches thick, frozen to fissure walls; less commonly small amounts of sulphides are mixed with gouge and decomposed wall rock.

SILURIAN AND OVEE MINES

These two mines connect and are conveniently described togethei. The Silurian is opened by two tunnels in the limestone, one of which is short and caved at the portal. The other, about parallel to the first, trends N. 27° W. for 45 feet, then N. 10° W. for 285 feet. In addition to these two tunnels, 370 feet of crosscuts and drifts could be seen in the summer of 1922. The Ovee tunnel is in the limestone and trends N. 17° W. for 60 feet, thence N. 15° E. 35 feet, thence N. 5° W. 220 feet, at which point it is caved. An old map shows 300 feet of development beyond this point. At the time of examination about 275 feet of short crosscuts from the main tunnel were accessible, most of which ended in caved stopes or raises, or were filled to the back.

The wall rock is "zebra" limestone. It is dolomitic and somewhat siliceous. In places it is altered to gray or brown jasperoid. The limestone is broken and faulted in both mines, but the larger faults are not well defined and can not be described. There appears to have been much shearing and brecciation with little net displacement. Most of the small faults strike northwest, and dip 15° to 24° northeast. A few strike northeast, and have high dips to the northwest or southeast. Remnants of ore occur irregularly in patches, vugs, and streaks, replacing the limestone in or near fault zones; the ore shows fairly thorough oxidation. In places the mineralization is scanty, and oxides of manganese or iron are mixed with much calcite, dolomite, quartz, and, rarely, barite.

Small shipments of high-grade ore made between 1916 and 1919 by George E. Bowland, one of the owners, ran 217 to 361 ounces silver per ton. In 1923 Frank Tetreault, a lessee, was shipping ore that ran 91 to 141 ounces silver per ton. Gold in the ore is rare and in very small amount.

HENRIETTA MINE

This mine is opened by a tunnel trending N. 5° E. for at least 260 feet in the upper part of the limestone. At the time of examination 460 feet of crosscuts could be seen, but the more remote workings were caved.

The ore was found near fault zones where there is evidence of considerable shearing, but where the net displacement could not be measured. The faults strike between north and northeast, and dip from 30° to 90°. In general the direction of dip is southeast. Remnants of ore in streaks and patches replacing the limestone show nearly thorough oxidation. No large stopes were seen in the workings that were still accessible. The wall rock is dolomitic "zebra" limestone containing a little siderite and vugs with quartz crystals. In places the limestone is very friable and breaks down to "dolomite sand."

LITTLE MAY MINE

The Little May is opened by a tunnel which trends N. 7° W. at least 165 feet in the limestone. At this point it is caved; consequently most of the mine could not be seen.

The wall rock is "zebra" limestone, spotted and streaked with limonite. The ore probably occurs in pockets and streaks as it does in other similar mines in the immediate neighborhood. Oxides of manganese and a little pyrite in all stages of alteration may be seen with limonite in and near fault zones.

FOSTER COMBINATION MINE

This mine is opened by a tunnel which trends N. 39° W. for 280 feet in the limestone. Much of the mine is caved and inaccessible, but about 700 feet of workings could be seen in 1923 when H. T. Wenger, who was leasing the property, was shipping about $300 worth of ore per month.

The wall rock is a ferriferous dolomitic limestone. In places .it is hard and crystalline, and shows "zebra" texture; elsewhere it is decomposed and friable, especially near the fault zones. All through the mine are evidences of disturbed ground, but relationships are, for the most part, concealed since the ore has been extracted near fault zones, and the old workings are caved or lagged up. Three northeastward-striking faults were seen. The dip, which could be measured in only one of these faults, is 70° southeast; the southeast wall shows vertical grooves. In other places the position of brecciated areas indicates that stresses were relieved by shearing along rather flat-lying zones under the porphyry. The ore occurs in patches, pockets, vugs, and streaks in and near fissures and faults. The streaks, which may trend in any direction, widen and make pockets of ore. As they are followed upward, in many places the streaks turn, flatten out under the porphyry or shale, get wider, and produce more and better ore. Much of the mineralization is the result of replacement of the wall rock by sulphides which later altered to oxides; there are also small vugs partly filled with oxides and altered sulphides. The chief minerals are pyrite, galena, sphalerite, barite, gypsum, and oxides of iron and manganese.

Mr. Wenger stated that his high-grade ore averaged 200 ounces silver and the low grade 9 to 15 ounces of silver per ton. The ore carries very little gold. The mine is equipped with electric lights and an electric hoist.

FIRST NATIONAL AND EIGHTY FOUR MINES

As these mines are similar in most respects and are connected, they are described together. Caved crosscuts or water prevented a complete examination of either mine.

The First National is opened by a tunnel which trends N. 10° W. in the limestone. An old map shows two tunnels on the Eighty Four claim, both trending about N. 20° W., but both are caved at the portals. At the time the map was made the First National had 575 feet of development still accessible, and the Eighty Four had 1,080 feet. Very little of the latter mine could be seen. The wall rock is dolomitic "zebra" limestone that carries siderite. The ore occurs as pyrite in all stages of alteration, and limonite in patches and streaks in and near ill-defined fault zones. Pyrite has replaced the wall rock and filled thin, discontinuous fissures.

The report of the Director of the Mint for 1887 gives the value of the output of the Eighty Four for that year as $1,991. In 1915 Dismant Brothers and Company shipped ten lots of ore from the same mine which netted them $4,825. This ore carried 47 to 73 ounces silver per ton, 0.7

to 9.2 per cent lead, and as much as 12 per cent zinc. One lot contained a little copper, and two lots ran .02 ounces gold per ton. The reports of the Director of the Mint give the value of the production from the First National for the years 1888-1891 as $4,359. Both mines have produced more ore than mentioned, but the records are not available.

WYOMING VALLEY MINE

The Wyoming is opened by two tunnels which trend southeast in the limestone. The wall rock is dolomitic limestone with the "zebra" texture developed in places, and contains vugs of dolomite, siderite, and quartz crystals. It is rarely replaced by jasperoid, but commonly decomposes to "dolomite sand." At the time of examination most of the workings were inaccessible because of water and caving. Much of the following information was furnished by Messrs J. M. and R. V. Dismant.

There are numerous faults and fissures in the mine, most of which strike northeast and are vertical or have high dips. Three faults were seen which strike northwest. The dip of any one fault may vary in amount and direction; it may dip northwest in the upper workings and southeast on the lower levels. A few ill-defined, flat-lying, northeastward-dipping faults were observed in the upper part of the limestone near the base of the porphyry, which seem to dip a little more than the beds. It may be that some of these are bedding faults. The fault zones vary in width from a few inches to 4 feet.

The ore, most of which was oxidized, occurred in faulted and fissured zones and in flat-lying streaks and pockets as a replacement of the limestone. Remnants of ore showed oxides of iron and rare streaks of pyrite, galena, and sphalerite. The largest bodies were found above or close to the fissures in the upper beds near or adjacent to the porphyry. These flat ore shoots had an average thickness of less than 18 inches and a maximum of 4 feet. In lateral extent they were very irregular, the largest measuring approximately 150 feet across its greatest dimension. The mineralization in the steeply dipping fissures and faults was not wide, and in places occurred only on one side of the break.

Through the courtesy of Mr. J. M. Dismant, who permitted the writer to examine the smelter settlement sheets, the following information concerning the output of the mine from March 11, 1914, to May 18, 1920, is presented:

Output -- 4,307 tons
Lead --- 11,528 pounds
Silver --- 212,822 ounces
Gold --- 13.06 ounces
Gross value ------------------------------------- $128,584.45
Average value per ton --------------------------------$29.85

LIBERTY MINE

This mine is opened by an incline which trends a little east of north in the porphyry. As most of the mine was filled with water at the time of examination Mr. J. M. Dismant kindly furnished the writer with most of the following.

The incline is driven 250 feet down a 30-degree slope, and for 69 feet

more on a 40-degree slope. West of the incline a drift in the limestone below the porphyry, on a northwest-striking vein, encountered small pockets
of high-grade ore—blende and wire silver. Before much ore was extracted
the flow of water became too great to handle, and the mine was temporari-
ly abandoned. In the summer of 1923 Mr. Dismant was considering in-
stalling a larger pump and reopening the property.

HORN SILVER MINE

In 1879 the Horn Silver mine was opened by William Greiner and
G. J. Da Lee, and is said to be the first mine operated in the immediate
vicinity of Red Cliff. The ore first produced ran about $3,000 per ton, and
the mine is credited with a considerable production of high-grade silver
ore. The silver is reported to have been chiefly combined in the mineral
cerargyrite, or horn silver, and associated with oxides of iron and manganese.
In 1920 the latest shipment was made from the old workings which are
now caved at the entrance.

The new Horn Silver mine, operated by lessees in 1923, has been
producing silver ore said to average $80 per ton. The mine is worked
through a tunnel and incline in the Leadville limestone. The ore is a replacement deposit in irregular streaks in broken limestone not far below
the porphyry. Bedding faults, seen in the tunnel walls at a lower position
in the limestone, have streaks of gouge from a fraction of an inch to a
foot thick; oxides of iron and manganese are found in the gouge. The
mine is equipped with a gasoline hoist and fan.

INDEX

PUBLICATIONS OF THE SURVEY.

* FIRST REPORT, 1908: Out of print.

BULLETIN 1, 1910: Geology of Monarch Mining Dist., Chaffee Co.

BULLETIN 2, 1910: Geology of Grayback Mining Dist., Costilla Co.

BULLETIN 3, 1912: Geology and Ore Deposits of Alma Dist., Park Co.

BULLETIN 4, 1912: Geology and Ore Deposits of Monarch and Tomichi Districts, Chaffee and Gunnison Counties.

BULLETIN 5, 1912: Part I, Geology, Rabbit Ears Region, Grand, Jackson, and Routt Counties; Part II, Permian or Permo-Carboniferous of Eastern Foothills of Rocky Mts. in Colorado.

* BULLETIN 6, 1912: Common Minerals and Rocks, Their Occurrence and Uses.

BULLETIN 7, 1914: Bibliography Colorado Geology and Mining Literature.

BULLETIN 8, 1915: Clays of Eastern Colorado.

BULLETIN 9, 1915: Bonanza District, Saguache County.

BULLETIN 10, 1916: Geology and Ore Deposits of Gold Brick District, Gunnison Co.

BULLETIN 11, 1921: Mineral Waters of Colorado.

BULLETIN 12, 1917: Common Rocks and Minerals, Their Occurrence and Uses.

BULLETIN 13, 1917: Platoro-Summitville Dist., Rio Grande and Conejos Co.

BULLETIN 14, 1919: Molybdenum Deposits of Colorado.

BULLETIN 15, 1919: Manganese Deposits of Colorado.

BULLETIN 16, 1920: Radium, Uranium and Vanadium Deposits of Southwestern Colorado.

BULLETIN 17, 1921: Twin Lakes District, Lake and Chaffee Counties.

BULLETIN 18, 1920: Fluorspar Deposits of Colorado.

BULLETIN 19, 1921: Foothills Formations of North Central Colorado; Cretaceous Formations of Northeastern Colorado.

BULLETIN 20, 1924: Preliminary Notes for Revision of Geologic Map of Eastern Colorado.

BULLETIN 21, 1921: Geology of the Ward Region, Boulder County.

BULLETIN 22, 1921: Mineral Resources of the Western Slope.

* BULLETIN 23, 1920: Some Anticlines of Routt County, Inc. with 24.

BULLETIN 24, 1921: Some Anticlines of Western Colorado.

BULLETIN 25, 1921: Oil Shales of Colorado.

BULLETIN 26, 1921: Preliminary Report, Underground Waters, S. E. Colorado.

BULLETIN 27, 1923: Part I, Underground Waters of La Junta Area; Part II, Underground Waters of Parts of Crowley and Otero Counties; Part III, Geology of Parts of Las Animas, Otero and Bent Counties.

BULLETIN 28, 1923: Oil and Water Resources, Parts of Delta and Mesa Co.

BULLETIN 29, 1924: Physiography of Colorado.

BULLETIN 30, 1924: Geology and Ore Deposits, Redcliff Mining District.

BULLETIN 31, 1924: The Como Mining District. Ready for printing.

REPORT 2, 1922. Geology of Line of Proposed Moffat Tunnel.

OIL MAP OF COLORADO: A reconnaissance map showing locations of seepages, anticlines, etc. $1.00.

TOPOGRAPHIC MAP OF COLORADO, 1913: Scale of eight miles to inch. Unmounted, $3.00; Mounted, $4.00; Mounted with sticks, $5.00.

GEOLOGIC MAP OF COLORADO, 1913: Presents by colors and patterns the geological formations of the state. Supply almost exhausted. Prices, $5.00, $6.00, and $7.00.

COAL MAP OF COLORADO: A reconnaissance map showing known areas underlain by coal in the state. $1.00.

STRATIGRAPHIC SECTIONS, EASTERN COLORADO: $1.00.

---

*Out of Print.

Address

COLORADO GEOLOGICAL SURVEY

Boulder, Colorado

This original publication came with additional plates and/or maps included in the document or the back pocket of the publication. Miningbooks.com has digitally scanned and formatted these plates and/or maps and put them on a CD ROM. Due to the printing and distribution process we use, some of our distributors and retailers may not have the capability to add this CD in with their drop-ship process. If you did not receive this CD Rom with your book we have made this CD available directly from our website www.miningbooks.com for you to purchase at a nominal cost in order to cover shipping , handling, and materials.

www.ingramcontent.com/pod-product-compliance
Lightning Source LLC
Chambersburg PA
CBHW060635210326
41520CB00010B/1619